AF564525

FARMERS OF FORTY CENTURIES

FARMERS OF FORTY CENTURIES

or Permanent Agriculture in China, Korea and Japan

F.H. King

Edited by
J.P. Bruce

First Published 1911
First LG Edition 2021

ISBN 978–93–83723–72–0

Published by
LG PUBLISHERS DISTRIBUTORS
49, Street No. 14, Pratap Nagar,
Mayur Vihar Phase I, Delhi 110091
lgpdist@gmail.com

Printed at
Sapra Brothers, Noida

CONTENTS

LIST OF ILLUSTRATIONS

PREFACE

BY DR. L. H. BAILEY

We have not yet gathered up the experience of mankind in the tilling of the earth; yet the tilling of the earth is the bottom condition of civilization. If we are to assemble all the forces and agencies that make for the final conquest of the planet, we must assuredly know how it is that all the peoples in all places have met the problem of producing their sustenance out of the soil.

We have had few great agricultural travellers and few books that describe the real and significant rural conditions. Of natural history travel we have had very much; and of accounts of sights and events perhaps we have had too many. There are, to be sure, famous books of study and travel in rural regions, and some of them, as Arthur Young's *Travels in France*, have touched social and political history; but for the most part, authorship of agricultural travel is yet undeveloped. The spirit of scientific inquiry must now be taken into this field, and all earth-conquest must be compared and the results given to the people that work.

This was the point of view from which I read Professor King's manuscript. It is the writing of a well-trained observer who went forth not to find diversion or to depict scenery and common wonders, but to study the actual conditions of life of agricultural peoples. We in North America are wont to think that we may instruct all the world in agriculture, because our agricultural wealth is great and our exports to less favoured peoples have been heavy; but this wealth is great because our soil is fertile and new, and in large acreage for every person. We have really only begun to farm well. The first condition of farming is to maintain fertility. This condition the oriental peoples have met, and they have solved it in their way. We may never adopt particular methods, but we can profit vastly by their experience. With the increase of personal wants in recent time, the newer countries may never reach such density of population as have Japan and China; but we must nevertheless learn the first lesson in the conservation of natural resources, which are the resources of the land. This is the message that Professor King brought home from the East.

This book on agriculture should have good effect in establishing understanding between the West and the East. If there could be such an interchange of courtesies and inquiries on these themes as is suggested by Professor King, as well as the interchange of athletics and diplomacy and commerce, the common productive people on both sides should gain much that they could use; and the results in amity should be incalculable.

It is a misfortune that Professor King could not have lived to write the concluding *Message of China and Japan to the World*. It would have been a careful and forceful summary of his study of eastern conditions. At the moment when the work was going to the printer, he was called suddenly to the endless journey and his travel here was left incomplete. But he bequeathed us a new piece of literature, to add to his standard writings on soils and on the applications of physics and devices to agriculture. Whatever he touched he illuminated.

L. H. BAILEY.

INTRODUCTION

A WORD of introduction is needed to place the reader at the best view-point from which to consider what is said in the following pages regarding the agricultural practices and customs of China, Korea and Japan. It should be borne in mind that the great factors which to-day characterize and determine the agricultural and other industrial operations of western nations were physical impossibilities to them as to all other peoples one hundred years ago.

It should be observed, too, that the United States as yet is a nation of but few people widely scattered over a broad virgin land with more than twenty acres to the support of every man, woman and child, while the people whose practices are to be considered are toiling in fields which have been tilled for more than three thousand years and who have scarcely more than two acres *per capita*,[1] more than one-half of which is uncultivable mountain land.

Again, the great movement of cargoes of feeding-stuffs and mineral fertilizers to western Europe and to the eastern United States began less than a century ago and has never been possible as a means of maintaining soil fertility in China, Korea or Japan, nor can it be continued indefinitely in either Europe or America. These importations are for the time making tolerable the waste of plant food materials through our modern systems of sewage disposal and other faulty practices; but the Mongolian races have held all such wastes, both urban and rural, and many others which we ignore, sacred to agriculture, applying them to their fields.

We are to consider some of the practices of a virile race of some 500 millions of people who have an unimpaired inheritance moving with the momentum acquired through 4,000 years; a people morally and intellectually strong, mechanically capable, who are awakening to a utilization of all the possibilities which science and invention during recent years have brought to western nations; and a people who have long dearly loved peace but who can and will fight in self-defence if compelled to do so.

[1] This figure was wrongly stated in the first edition as one acre, owing to a mistake in confusing the area of cultivated land with total area.

We had long desired to stand face to face with Chinese and Japanese farmers, the oldest farmers in the world; to walk through their fields and to learn by seeing some of their methods, appliances and practices which centuries of stress and experience have led them to adopt. We desired to learn how it is possible, after twenty and perhaps thirty or even forty centuries, for their soils to be made to produce sufficiently for the maintenance of such dense populations as are living now in these three countries. We have now had this opportunity and almost every day we were instructed in the ways and extent to which these nations for centuries have been conserving and utilizing their natural resources; we were surprised at the magnitude of the returns they are getting from their fields, and amazed at the amount of efficient human labour cheerfully given for a daily wage of five cents and their food, or for fifteen cents, United States currency, without food.

The three main islands of Japan in 1907 had a population of 46,977,003 maintained on 20,000 square miles of cultivated field. This is at the rate of more than three people to each acre, and of 2,349 to each square mile; and yet the total agricultural imports into Japan in 1907 exceeded the agricultural exports by less than one dollar *per capita*. If the cultivated land of Holland is estimated at but one-third of her total area, the density of her population in 1905 was, on this basis, less than one-third that of Japan in her three main islands. At the same time Japan is feeding 69 horses and 56 cattle, nearly all labouring animals, to each square mile of cultivated field, while in the same area on an average we in the United States were feeding in 1900 not more than 30 horses and mules, our labouring animals.

As coarse food transformers Japan was maintaining 16,500,000 domestic fowl, 825 per square mile, but only one for almost three of her people. We were maintaining, in 1900, 250,600,000 poultry, but only 387 per square mile of cultivated field and yet more than three for each person. Japan's coarse food transformers in the form of swine, goats and sheep aggregated but 13 to the square mile and provided but one of these units for each 180 of her people; while in the United States in 1900 there were being maintained, as transformers of grass and coarse grain into meat and milk, 95 cattle, 99 sheep and 72 swine per each square mile of improved farms. In this reckoning each of the cattle should be counted as the equivalent of perhaps five of the sheep and swine, for the transforming power of the dairy cow is high. On this basis we are maintaining at the rate of more than 646 of the Japanese units per square mile, and more than five of these to every man, woman and child, instead of one to every 180 of the population, as is the case in Japan.

Correspondingly accurate statistics are not accessible for China, but in the Shantung province we talked with a farmer having 12 in his family who kept one donkey, one cow, both exclusively labouring animals, and two pigs on 2·5 acres of cultivated land where he grew wheat, millet, sweet potatoes and beans. Here is a density of population equal to 3,072 people, 256 donkeys, 256 cattle and 512 swine per square mile. In another instance where the holding was one and two-thirds acres the farmer had 10 in his family and was maintaining one donkey and one pig, showing for this farm land a maintenance capacity of 3,840 people, 384 donkeys and 384 pigs to the square mile; or 240 people, 24 donkeys and 24 pigs to one of our forty-acre farms which our farmers regard too small for a single family. The average of seven Chinese holdings which we visited and where we obtained similar data indicates a maintenance capacity of 1,783 people, 212 cattle or donkeys and 399 swine – 1,995 consumers and 399 rough food transformers per square mile of farm land. These statements for China represent strictly rural populations. The rural population of the United States in 1900 was placed at 61 per square mile of improved farm land and there were 30 horses and mules. In Japan the rural population had a density in 1907 of 1,922 per square mile, and in addition there were 125 horses and cattle.

The population of the large island of Chungming in the mouth of the Yangtse River, with an area of 270 square miles, possessed, according to the official census of 1902, a density of 3,700 per square mile; and yet, as there is but one large city on the island, the population is largely rural.

It could not be other than a matter of the highest industrial, educational and social importance to any nation if it could be furnished with a full and accurate account of all those conditions which have made it possible for such dense populations to be maintained upon the products of Chinese, Korean and Japanese soils. Many of the steps, phases and practices through which this evolution has passed are irrevocably buried in the past, but such remarkable maintenance efficiency attained centuries ago and projected into the present with little apparent decadence merits the most profound study. Living as we are in the morning of a century of transition from isolated to cosmopolitan national life, when profound readjustments, industrial, educational and social, must result, such an investigation cannot be made too soon. It is high time for each nation to study other nations and, by mutual agreement and co-operative effort, the results of such studies should be made available for the rest, so that all may become co-ordinate and mutually helpful component factors in the world's progress.

One very appropriate and immensely helpful means for attacking this problem would be for the higher educational institutions of all nations, instead of exchanging courtesies through their baseball teams, to send select bodies of their best students, under competent leadership and by international agreement, both east and west, to study specific problems. Such a movement, well conceived and directed, manned by the most capable young men, would spread broadcast a body of important knowledge which would contribute immensely to world peace and world progress. If some broad plan of international effort such as is here suggested were organized, the expense of maintenance might well be met by diverting so much as is needful from the large sums set aside for the expansion of navies; for such steps as these, taken in the interests of world uplift and world peace, could not fail to be more efficacious and less expensive than increase in fighting equipment. It would cultivate the spirit of pulling together and of a square deal rather than one of holding aloof and of striving to gain unneighbourly advantage.

Many factors and conditions conspire to give to the farms and farmers of the Far East their high maintenance efficiency, and some of these may be succinctly stated. The portions of China, Korea and Japan, where dense populations have developed and are being maintained, occupy exceptionally favourable geographic positions so far as these influence agricultural production.

Canton in the south of China has the latitude of Havana, Cuba, while Mukden in Manchuria, and northern Honshu in Japan, are only as far north as New York city, Chicago and northern California. The United States lies mainly between 50° and 30° of latitude, while these three countries lie between 40° and 20°, some 700 miles further south. This difference of position, giving them longer seasons, has made it possible for them to devise systems of agriculture whereby they grow two, three and even four crops on the same piece of ground each year. In southern China, in Formosa and in parts of Japan two crops of rice are grown; in the Chekiang province there may be a crop of rape, of wheat or barley or of windsor beans or clover which is followed in midsummer by another of cotton or of rice. In the Shantung province wheat or barley in the winter and spring may be followed in summer by large or small millet, sweet potatoes, soy beans or peanuts. At Tientsin, 39° north, in the latitude of Cincinnati, Indianapolis, and Springfield, Illinois, we talked with a farmer who followed his crop of wheat on his small holding with one of onions and the onions with cabbage, realizing from the three crops at the rate of $163, gold, per acre; and with another who planted Irish potatoes at the earliest opportunity in the spring, marketing them when small, then followed these with radishes, and the radishes with cabbage, realizing from the three crops at the rate of $203 per acre.

Nearly 500,000,000 people are being maintained, chiefly upon the products of an area smaller than the improved farm lands of the United States. Complete a square on the lines drawn from Chicago southward to the Gulf and westward across Kansas, and there will be enclosed an area greater than the cultivated fields of China, Korea and Japan, from which five times our present population are fed.

The rainfall in these countries is not only larger than that even in our Atlantic and Gulf States, but it falls more exclusively during the summer season when its efficiency in crop production may be highest. South China has a rainfall of some 80 inches

In Shantung there are not two crops every year, but three crops in two years. The usual rotation is: Wheat in spring, beans in October, after which millet is sown and gathered in the following September. [Ed.]

with little of it during the winter, while in our southern states the rainfall is nearer 60 inches with less than one-half of it between June and September. Along a line drawn from Lake Superior through central Texas the yearly precipitation is about 30 inches, but only 16 inches of this falls during the months of May to September; while in the Shantung province, China, with an annual rainfall of little more than 24 inches, 17 of these fall during the months designated and most of this in July and August. When it is stated that under the best tillage and with no loss of water through percolation, most of our agricultural crops require 300 to 600 tons of water for each ton of dry substance brought to maturity, it can be readily understood that the right amount of available moisture, coming at the proper time, must be one of the prime factors of a high maintenance capacity for any soil, and hence that in the Far East, with their intensive methods, it is possible to make their soils yield large returns.

The selection of rice and of the millets as the great staple food crops of the three nations specified, and the systems of agriculture they have evolved so as to realize the largest possible yield from them, are to us remarkable, and indicate a grasp of essential principles which may well cause western nations to pause and reflect.

Notwithstanding the large and favourable rainfall of these countries, the people have in each case selected the one crop which permits them to utilize not only practically the entire amount of rain which falls upon their fields, but in addition enormous volumes of the run-off from adjacent uncultivable mountain country. Wherever paddy fields are practicable there rice is grown. In the three main islands of Japan 56 per cent of the cultivated fields, 11,000 square miles, is laid out for rice-growing and is maintained under water from transplanting to near harvest time, after which the land is allowed to dry, to be devoted to dry land crops during the balance of the year, where the season permits.

To anyone who studies the agricultural methods of the Far East in the field it is evident that these people, centuries ago, came to appreciate the value of water in crop production as no other nations have. They have adapted conditions to crops and crops to conditions to such a pitch that in rice they have produced a cereal which permits the most intense fertilization and at the same time ensures the maximum yields against both drought and flood. With the practice of western nations in all humid climates, no matter how completely and highly we fertilize, in more years than not, yields are reduced by a deficiency or an excess of water.

It is difficult to convey, by word or map, an adequate conception of the magnitude of the systems of canalization which contribute primarily to rice culture. A conservative estimate would place the miles of canals in China at fully 200,000, and there are probably more miles of canal in China, Korea and Japan than there are miles of railroad in the United States. China alone has as many acres in rice each year as the United States has in wheat, and her annual product is more than double and probably threefold our annual wheat crop, and yet the whole of the rice area produces at least one and sometimes two other crops each year.

The selection of the quick-maturing, drought-resisting millets as the great staple food crops to be grown wherever water is not available for irrigation, and the almost universal planting in hills or drills, and so making possible the utilization of earth mulches in conserving soil moisture, has enabled these people for centuries past to secure maximum returns in seasons of drought and in places where the rainfall is small. The millets thrive in the hot summer climates; they survive when the available soil moisture is reduced to a low limit, and they grow vigorously when the heavy rains come. Thus we find that in the Far East, with more rainfall and a better distribution of it than occurs in the United States, and with warmer, longer seasons, these people have with rare wisdom combined both irrigation and dry farming methods to an extent and with an intensity far beyond anything our people have ever dreamed of, in order that they might maintain these dense populations.

Notwithstanding the fact that in each of these countries the soils are naturally more than ordinarily deep, inherently fertile and enduring, judicious and rational methods of fertilization are everywhere practised; but not until recent years, and only in Japan, have mineral commercial fertilizers been used. For centuries, however, the canals, streams and the sea have been made to contribute toward the fertilization of cultivated fields, and these contributions in the aggregate have been large. In China, in Korea and in Japan all but the inaccessible portions of their vast extent of mountain and hill lands have long been taxed to their full capacity for fuel, timber, and herbage, for green manure and compost material; and the ash of practically all the fuel and of all the timber used in the home finds its way ultimately to the fields as fertilizer.

In China enormous quantities of canal mud are applied to the fields, sometimes at the rate of even 70 and more tons per acre. So, too, where there are no canals, both soil and subsoil are carried into the villages and there they are, at the expense of great labour, composted with organic refuse, then dried and pulverized, and finally carried back to the fields to be used as home-made fertilizers. Manure of all kinds, human and animal, is religiously saved and applied to the fields in a manner which secures an efficiency far above our own practices. Statistics obtained through the Bureau of Agriculture, Japan, place the amount of human waste in that country in 1908 at 23,950,295 tons, or 1·75 tons per acre of her cultivated land. The International Concession of the city of Shanghai, in 1908, sold to a Chinese contractor the privilege of entering residences and public places early in the morning of each day in the year and removing the night soil, at a price of more than $31,000, gold, for 78,000 tons of waste. We expend much larger sums in throwing all this away!

Japan's production of fertilizing material, regularly prepared and applied to the land annually, amounts to more than 4·5 tons per acre of cultivated field exclusive of the commercial fertilizers purchased. Between Shanhaikwan and Mukden in Manchuria we passed, on June 18th, thousands of tons of the dry highly nitrified compost soil recently carried into the fields and laid down in piles where it was waiting to be 'fed to the crops.'

It was not until 1888, and then after a prolonged war of more than thirty years, generalled by the best scientists of all Europe, that it was finally conceded as demonstrated that leguminous plants acting as hosts for lower organisms living on their roots are largely responsible for the maintenance of soil nitrogen, drawing it directly from the air to which it is returned through the processes of decay. But centuries of practice had taught the Far East farmers that the culture and use of these crops are essential to enduring fertility, and so in each of the three countries with which we are dealing the extensive growing of legumes in rotation with other crops, for the express purpose of fertilizing the soil, is one of their old, fixed practices.

Just before, or immediately after, the rice crop is harvested, fields are often sowed to 'clover' (*Astragalus sinicus*) which is allowed to grow until near the next transplanting time when it is either turned under directly, or more often stacked along the canals and saturated with soft mud dipped from the bottom of the canal. After fermenting twenty or thirty days it is applied to the field. And so it is literally true that these old-world farmers whom we regard as ignorant, have long included legumes in their crop rotation, as indispensable to profitable agriculture.

Time is a function of every life-process as it is of every physical, chemical and mental reaction. The husbandman is an industrial biologist and as such is compelled to shape his operations so as to conform with the time requirements of his crops. The oriental farmer is a time economizer beyond all others. He utilizes the first and last minute and all that are between. The foreigner speaks of the Chinese as people who always take their time, are never in a fret, and never in a hurry. This is quite true, and they are so for the reason that they are a people who definitely set their faces toward the future and lead time by the forelock. They have long realized that much time is required to transform organic matter into forms available for plant food, and, although they are the heaviest users in the world, the largest portion of this organic matter is predigested with soil or subsoil before it is applied to their fields. This is at an enormous cost of human time and labour, but it practically lengthens their growing season and enables them to adopt a system of multiple cropping which would not otherwise be possible. By planting in hills and rows with intertillage it is very common to see three crops growing upon the same field at one time, but in different stages of maturity – one nearly ready to harvest, one just coming up, and the third at the stage when it is drawing most heavily upon the soil. By such practice, with heavy fertilization supplemented by irrigation when needful, the soil is made to do full duty throughout the growing season.

Further, notwithstanding the enormous acreage of rice planted each year, it is all in the first instance set in hills and later transplanted. By this method, the farmer saves in all ways – with the single exception of human labour, which is the one thing they have in excess. By thoroughly preparing the seed-bed, fertilizing highly and giving the most careful attention, he is able to grow on one acre, during 30 to 50 days, enough plants to occupy ten acres, and in the meantime on the remaining nine acres other crops are maturing. After these are harvested the fields are prepared to receive the rice which by this time is ready for transplanting. So that in effect this interval of time is added to the growing season of the rice crops.

Silk culture is a great and, in some ways, one of the most remarkable industries of the Orient. Remarkable for its magnitude; for having had its birthplace apparently in oldest China at least 2,700 years B.C.; for having grown out of the domestication of a wild insect of the woods; and for having lived through more than 4,000 years, expanding to such an extent that now as much as a million-dollar cargo of the product has been laid down on our western coast and rushed by special fast express to the east for the Christmas trade.

A low estimate of China's production of raw silk would be 120,000,000 pounds annually, and this with the output of Japan, Korea and a small area of southern Manchuria, would probably exceed 150,000,000 pounds annually, representing a total value of perhaps $700,000,000, equalling in value the wheat crop of the United States, but produced on less than one-eighth the area of our wheat fields.

The cultivation of tea in China and Japan is another of the great industries of these nations, taking rank with that of sericulture, if not above it, in the important part it plays in the welfare of the people. There is little reason to doubt that this industry has its foundation in the need of something to render boiled water palatable for drinking purposes. The drinking of boiled water is universally adopted in these countries as an individually available and thoroughly efficient safeguard against that class of deadly disease germs which so far it has been impossible to exclude from the drinking water in its natural state of any densely peopled country.

Judged by the success of the most thorough sanitary measures instituted up till now, and taking into consideration inherent difficulties which must increase enormously with increasing populations, it appears inevitable that modern methods must ultimately fail in sanitary efficiency and that absolute safety can be secured only in some manner having the equivalent effect of boiling drinking water, long ago adopted by the Mongolian races.

In the year 1907 Japan had 124,482 acres of land in tea plantations, producing 60,877,975 pounds of cured tea. In China the volume annually produced is much larger than that of Japan, 40,000,000 pounds going annually to Tibet alone from the Szechwan province. The direct export to foreign countries was, in 1905, 176,027,255 pounds, and in 1906 it was 180,271,000, so that their annual export must exceed 200,000,000 pounds with a total annual output more than double this amount of cured tea.

But above any other factor, and perhaps greater than all of them combined in contributing to the high maintenance-efficiency attained in these countries, must be placed the standard of living to which the industrial classes have been compelled to adjust themselves, combined with the most rigorous economy which they practise along every line of effort and of living.

Almost every foot of land is made to contribute material for food, fuel or fabric. Everything which can be made edible serves as food for man or domestic animals. Whatever cannot be eaten or worn is used for fuel. The wastes of the body, of fuel and of fabric are taken back to the field; before doing so they are housed against waste from weather, intelligently compounded and patiently worked at through one, three or even six months, in order to bring them into the most efficient form to serve as manure for the soil, or as feed for the crop. It seems to be a golden rule with these industrious people, or if not golden, then an inviolable one, that whenever an extra hour or day of labour can promise even a little larger return, it must be given, and nothing be permitted to cancel the obligation or defer its execution.

I

FIRST GLIMPSES OF JAPAN

We left the United States from Seattle for Shanghai, China, sailing by the northern route, February 2nd, reaching Yokohama February 19th and Shanghai March 1st. It was our aim throughout the journey to keep in close contact with the field and crop problems and to converse personally, through interpreters or otherwise, with the farmers, gardeners and fruit growers themselves. We have taken pains in many cases to visit the same fields or the same region two, three or more times at different intervals during the season in order to observe different phases of the same cultural or fertilization methods as they changed or varied with the season.

Our first near view of Japan came in the early morning of February 19th. The high rounded hills were clothed neither in the dense dark forest green of Washington and Vancouver, left sixteen days before, nor yet in the brilliant emerald such as Ireland's hills in June fling in unparalleled greeting to passengers surfeited with the dull grey of the rolling ocean. This lack of strong forest growth, and even of shrubs and heavy herbage, on hills covered with deep soil, was our first great surprise.

To the southward around the point, after turning northward into the deep bay, similar conditions prevailed, and at ten o'clock we stood off Uraga where Commodore Perry anchored on July 8th, 1853, bearing to the Shogun President Fillmore's letter which opened the doors of Japan to the commerce of the world. As the *Tosa Maru* drew alongside the pier at Yokohama it was raining hard and an army attired after the manner of Robinson Crusoe, dressed as seen in Fig. 1, awaited us ready to carry us to the Customs house and beyond for one, two, three or five cents.

Through the kindness of Captain Harrison of the *Tosa Maru* in calling an interpreter by wireless to meet the steamer, it was possible to utilize the entire interval of our stay in Yokohama to the best advantage in the fields and gardens spread over the eighteen miles of plain extending to Tokyo. This wonderfully fertile and highly tilled district was traversed by both electric

Fig. 1. – Rainy weather costume, as worn in Japan and typical of those used under similar conditions in both Korea and China. The picture shows a group of Japanese rice field labourers with their most common tools.

tram and railway lines, each line running many trains and making frequent stops; so that almost any point could be readily and easily reached.

We had left home in a memorable storm of snow, sleet and rain, which cut out of service telegraph and telephone lines over a large part of the United States; we had seen nothing on the journey which could suggest a warm soil and green fields, hence our surprise was great to find the jinricksha men with bare feet, and legs naked to the thighs; and greater still when we found, before we were outside the city limits, that our electric tram was running between fields and gardens green with wheat, barley, onions, carrots, cabbage and other vegetables. We were rushing through the Orient with everything outside the car so strange and different from home that the shock came like a thunderbolt out of a clear sky.

In the car every man except myself and one other was smoking tobacco and that other was inhaling camphor through an ivory mouthpiece resembling a cigar holder closed at the end. The streets were muddy from the rain and everybody Japanese was wearing rainy-day wooden shoes, the soles carried three to four inches above the ground by two cross blocks. A mother, with a baby on her back, and a daughter of sixteen years came into the car. Notwithstanding her high shoes the mother had dipped one toe into the mud. Seated, she slipped her foot off. Without evident instructions the pretty black-eyed, glossy-haired, red-lipped lass, with cheeks made rosy, picked up the shoe, withdrew a piece of white tissue paper from the great pocket in her sleeve, deftly cleaned the otherwise spotless white cloth sock and then the shoe, threw the paper on the floor, looked to see that her fingers were not soiled, then set the shoe at her mother's foot, which found its place without effort or glance.

Everything here was strange and the scenes shifted with the speed of the wildest dream. Now it was driving piles for the foundation of a bridge. A tripod of poles was erected above the pile and from it hung a pulley. Over the pulley passed a rope from the driving weight and from its end at the pulley ten cords extended to the ground. In a circle at the foot of the tripod stood ten agile Japanese women. They were the hoisting engine. They chanted in perfect rhythm, hauled and stepped, dropped the weight and hoisted again, making up for heavier hammer and higher drop by more blows per minute. When we reached Shanghai we saw the pile driver being worked from above. Fourteen Chinese men stood upon a raised staging, each with a separate cord passing direct from the hand to the weight below. A concerted, half-musical chant, modulated to relieve monotony, kept all hands together. What did the operation of this machine cost? Thirteen cents, gold, per man per day, which covered fuel and lubricant, both automatically served. Two additional men managed the piles, two directed the hammer, eighteen manned the outfit. Two dollars and thirty-four cents per day covered fuel, superintendence and repairs. There was almost no capital invested in machinery. Men were plentiful. Rice was the fuel, cooked without salt, boiled stiff, reinforced with a bit of pork or fish, appetized with salted cabbage or turnip, and perhaps two or three of forty and more other vegetable relishes. Are these men strong and happy? They certainly are strong. They are steadily increasing their millions, and as one stood and watched them at their work their faces were often wreathed in smiles and wore what seemed to be a look of satisfaction and contentment.

Among the most common early morning sights on the journey from Yokohama to Tokyo were the loads of night soil carried on the shoulders of men and on the backs of animals, but most commonly on strong carts drawn by men. Each cart carried from six to ten tightly covered wooden containers holding forty, sixty or more pounds each. Strange as it may seem, there are not to-day, and apparently never have been, even in the largest and oldest cities of Japan, China or Korea, anything corresponding to the hydraulic systems of sewage disposal used now by western nations. Provision is made for the removal of storm waters, but when I asked my interpreter if it was not the custom of the city during the winter months to discharge its night soil into the sea, as a quicker and cheaper mode of disposal, his reply came quick and sharp, 'No, that would be waste. We throw nothing away. It is worth too much money.' In such public places as railway stations provision is made for saving, not for wasting, and even along the country roads screens invite the traveller to stop, primarily for profit to the owner rather than for personal convenience.

Between Yokohama and Tokyo, along the electric car line

FIG. 2. – Method of drying seaweed used for food. The small black squares on the larger light ones are the seaweed. Skewers pin the squares of matting against the long screens, six of which are shown in parallel series.

FIG. 3. – Section of shallow sea bottom planted with brushwood on which the edible seaweeds attach themselves and grow.

and not far distant from the seashore, there were to be seen in February very many long, fence-high screens built of rice straw, closely tied together and supported on bamboo poles carried upon posts of wood set in the ground. They extended east and west, and were strongly inclined towards the north. These screens, set in parallel series of five to ten or more in number and several hundred feet long, were used for the purpose of drying varieties of delicate seaweed, spread out in the manner shown in Fig. 2.

The seaweed is first spread upon separate 10 by 12 inch straw mats, forming a thin layer 7 by 8 inches. These mats are held by means of wooden skewers forced through the body of the screen, exposing the seaweed to the direct sunshine. When dry the rectangles of seaweed are piled in bundles an inch thick, cut in two, so as to form packages 4 by 7 inches. The packages are then neatly tied together and exposed for sale as soup stock and for other purposes.

To obtain this seaweed from the ocean, small shrubs and the limbs of trees are set up in the bottom of shallow water (Fig. 3). To these limbs the seaweeds become attached, grow to maturity and are then gathered by hand. By this method of culture large amounts of important food-stuff are grown for the support of the people on areas otherwise wholly unproductive.

Another rural feature, best shown by photograph taken in February, is the method of training pear orchards in Japan. Their limbs are tied down upon horizontal overhead trellises at such a height that a man can readily walk erect underneath and easily reach the fruit with the hand while standing upon the ground. Pear orchards thus form arbours of greater or less size, the trees being set in quincunx order about 12 feet apart in and between the rows. Bamboo poles are used overhead, carried on posts of the same material 1·5 to 2·5 inches in diameter, to which they are tied (Fig. 4).

The limbs of the pear trees are trained strictly in one plane. They are tied down and those not desired are pruned out. As a result the ground beneath is completely shaded and every pear is within reach. The accessibility of the fruit is a great convenience when it becomes necessary to tie paper bags over every pear in order to protect it from insects (Figs. 5 and 6). The orchard ground is kept free from weeds and not infrequently covered with a layer

FIG. 4. – Looking down upon an extensive pear orchard, the limbs of which are trained horizontally, forming an arbour completely shading the ground when in leaf, and placing all of the fruit within reach of the hand from beneath.

FIG. 5. – Pear trees at Akashi Experiment Station, Japan. Pears protected by paper bags. Special form of pruning advised by Prof. Ono, standing on the left, with Prof. Tokito. The trees branch below rather than at the level of the trellis.

of rice or other straw, which is extensively used in Japan as a ground cover with various crops. When so used it is carefully laid in handfuls, the straws being kept parallel as when harvested.

To the traveller coming from a country of 160-acre farms, with roads four rods wide; of cities with broad streets and residences, with green lawns and ample back yards; a country too where the

Fig. 6. – Low-branching pear orchard with pears protected by paper bags, at Akashi Experiment Station, Japan.

cemeteries are large and beautiful parks, the contrast presented by the over-crowding which he notices in the very first days of travel in these old countries forces itself upon his attention. The cities are over-crowded with houses and shops, and these with people and wares; the country is over-crowded with fields and the fields with crops; while in Japan the over-crowding is greatest of all in the cemeteries, where the gravestones almost touch each other; and in the surrounding country dwellings, gardens or rice fields contest the tiny allotted areas too closely to leave even footpaths between.

Fig. 7. – Street in a Hakone country village. The general absence of old forest growth on the hills is characteristic of much of Japan.

FIG. 8.—Chinese country village lining both sides of a canal. Section one-third of a mile long between two bridges, where in three rows of houses live 240 families.

Unless recently modified through foreign influence, the streets of villages and cities are narrow, as seen in Fig. 7, where the street, narrow as it is, is broader than most. This is a village in the Hakone district on a beautiful lake of the same name, where stands an Imperial summer palace, seen near the centre of the view on a hill across the lake. The roofs of the houses here are typical of the neat careful thatching with rice straw, very generally adopted in place of tile for the country villages.

In the canalized regions of China the country villages crowd both banks of a canal, as is the case in Fig. 8. Here, too, there is often a single street, very narrow, very crowded and very busy. Stone steps lead from the houses down into the water where clothing, vegetables, rice and what not are conveniently washed. In this particular village two rows of houses stand on one side of the canal separated by a very narrow street, and a single row on the other. Between the bridge, where the camera was exposed, and another bridge barely discernible in the background, a third of a mile distant, we counted upon one side, walking along the narrow street, 80 houses, each with its family, usually of three generations and often of four. Thus in the narrow strip, 154 feet broad, including 16 feet of street and 30 feet of canal, with its three lines of houses, lived no less than 240 families and more than 1,200 and probably nearer 2,000 people.

When we turn to the crowding of fields in the country nothing except seeing can reveal the fact so forcibly as the landscapes shown in Figs, 9, 10 and 11, one in Japan, one in Korea, and one in China. The latter is a scene not far from Nanking, looking from the hills across the fields to the broad Yangtse-kiang, barely discernible as a band of light along the horizon.

The average area of the rice field in Japan is less than 5 square rods and that of her upland fields only about 20. In the case of the rice fields the small size is necessitated partly by the requirement of holding water on the sloping sides of the valley (Fig. 9). These small areas do not represent the amount of land worked by one family, the average for Japan being more nearly 2·5 acres. But the lands worked by one family are seldom contiguous, they may even be widely scattered and very often rented.

The people generally live in villages, going often considerable distances to their work. Recognizing the great disadvantage of

FIG. 9.—Closely crowded fields of rice in Japan, each field filled with water and the rice recently transplanted.

Fig. 10. – Landscape in Korea, showing subdivision of the valley surface into small irregular fields separated only by narrow, low ridges of earth scarcely more than a foot wide and a foot high. The centre field is planted with rice, fields on the right are ploughed and watered but not fitted, the ridged field on the left is watered but not ploughed.

FIG. 11. – Landscape of rice fields in China. Fields in the foreground still covered with winter crops, but when harvested, to be planted with rice. White areas flooded with water and fitting for rice. Yangtse River near horizon.

scattered holdings broken into such small areas, the Japanese Government has passed laws for the adjustment of farm lands, which have been in force since 1900. They provide for the exchange of lands; for changing boundaries; for changing or abolishing roads, embankments, ridges or canals and for alterations in irrigation and drainage which would ensure that there would be larger areas and that the channels and roads would be straightened and made less numerous and less wasteful of time, labour and land. Up to 1907 Japan had issued permits for the readjustment of over 240,000 acres, and Fig. 12 is a landscape in one of these readjusted districts. To provide capable experts for planning and supervising these changes, the Government in 1905 entrusted the training of men to the higher agricultural school belonging to the Dai Nippon Agricultural Association, and since 1906 the Agricultural College and the Kogyokusha have undertaken the same task. Now there are men sufficient to push the work as rapidly as desired.

It may be remembered, too, as showing how, along other fundamental lines, Japan is taking effective steps to improve the condition of her people, that she already has her Imperial highways extending from one province to another; her prefectural roads which connect the cities and villages within the prefecture; and those which serve the farms and villages. Each of the three systems of roads is maintained by a specific tax levied for the purpose which is expended under proper supervision, a designated section of road being kept in repair through the year by a specially appointed crew, as is the practice in railroad maintenance. The result is, Japan has roads maintained in excellent condition, always narrow, sacrificing the minimum of land, and everywhere without fences.

That the fields are crowded with crops and all available land is made to do full duty is evident in Fig. 13, where even the narrow dividing ridges which retain the water are bearing a heavy crop of soy beans; and where may be seen a narrow pear orchard standing on the very slightest rise of ground, not a foot above the water.

How closely the ground itself may be crowded with plants is seen in Fig. 14, where a young peach orchard, whose tree-tops were 6 feet through, planted in rows 22 feet apart, had also

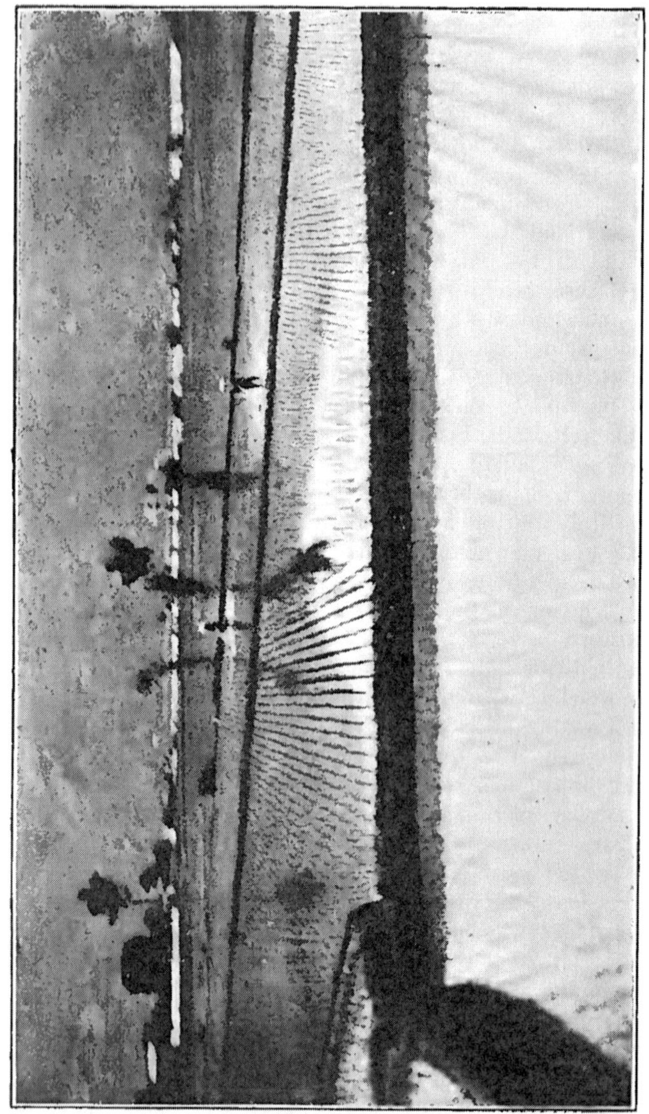

FIG. 12. – Landscape in one of the readjusted districts in Japan where division lines between paddy fields have been straightened. Men using new rice-weeding cultivators.

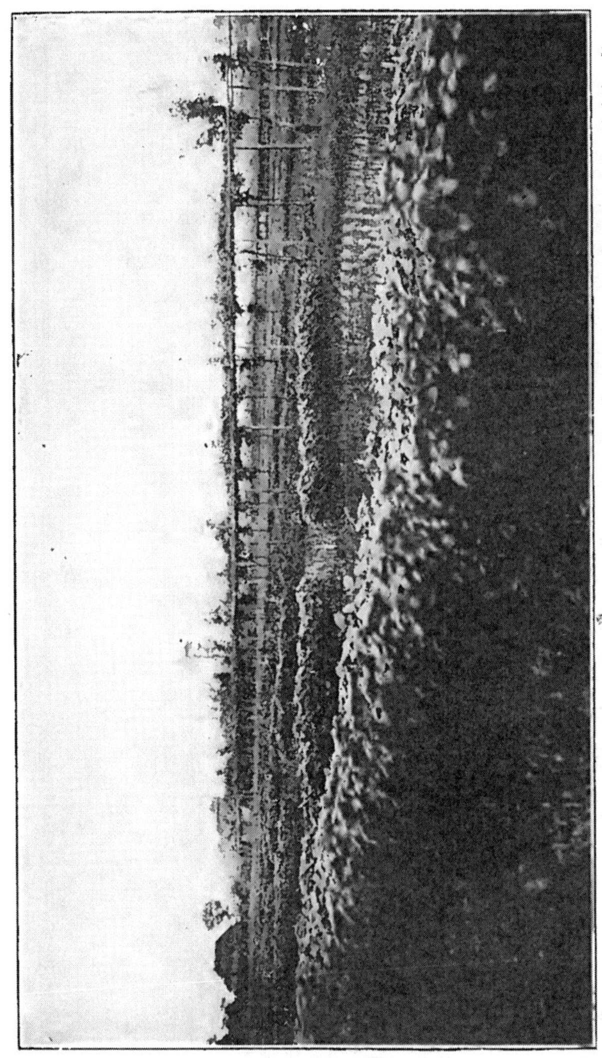

Fig. 13. – The entire field completely occupied by crops, rendering effective service. Soy beans on the dividing lines, rice in the paddy fields, pear orchard on the narrow raised ridge.

Fig. 14. – Young peach orchard doing intense duty as a market garden, growing peas, cabbage and windsor beans.

ten rows of cabbage, two rows of large windsor beans and a row of garden peas. Thirteen rows of vegetables in 22 feet, all luxuriant and strong! Note the judgment shown in placing the tallest plants, needing the most sun, in the centre between the trees.

But these old-world people, accustomed to crowding and to being crowded, and long ago capable of making four blades of grass grow where Nature grew but one, have also learned how to double the acreage where a crop needs more elbow room than it does standing room, as seen in Fig. 15. This man's garden had an area of but 63 by 68 feet, and two square rods of this was held sacred to the family grave mound. Yet his statement of yields, number of crops and prices made his earning $100 a year on less than one-tenth of an acre. His crop of cucumbers on less than ·06 of an acre would bring him $20. He had already sold $5 worth of greens and a second crop would follow the cucumbers. He had just irrigated his garden from an adjoining canal, using a foot-power pump, and stated that until it rained he would repeat the watering once a week.

On February 21st the *Tosa Maru* left Yokohama for Kobe at schedule time. On reaching Kobe we transferred to the *Yamaguchi Maru*, which sailed the following morning, to shorten the time of reaching Shanghai. This left but an afternoon for a trip into the country between Kobe and Osaka, where we found, if possible, even higher and more intensive culture practices than on the Tokyo plain, there being less land not carrying a winter crop. Fig. 17 shows how closely the crops crowd the houses and shops. Here were many cement-lined cisterns or sheltered reservoirs for collecting manures and preparing fertilizers, and the appearance of both soil and crops showed in a marked manner to what advantage. We passed a garden of nearly an acre entirely devoted to English violets just coming into full bloom. They were grown in long parallel east and west beds about 3 feet wide. On the north edge of each bed was erected a rice-straw screen 4 feet high which inclined to the south, overhanging the bed at an angle of some thirty-five degrees, thus forming a sort of oven-tent which reflected the sun, broke the force of the wind and checked the loss of heat absorbed by the soil.

The voyage from Kobe to Moji was made between 10 in the morning, February 24th, and 5.30 p.m. of February 25th, over a

Fig. 15. – Increasing the available surface of the field so that double the number of plants may occupy the ground. A row of cucumbers on opposite sides of each trellis will cover its surface.

FIG. 16. – Aged Chinese farmer in winter dress.

Fig. 17. – Newly started gardens crowded close about houses and shops between Kobe and Osaka.

quiet sea. We left Moji again in the early morning, and late in the evening of the same day entered the beautiful harbour of Nagasaki, all on board waiting until morning for a launch to go ashore. We were to sail again at noon, so available time for observation was short and we set out in a ricksha at once for our first near view of terraced gardening on the steep hillsides in Japan. Both going and returning our course led through streets paved with long, thick and narrow stone blocks, with deep open gutters on one or both sides close to the houses, into which waste water was emptied and through which the storm waters found their way to the sea. Few of these streets were more than 12 feet wide, and close watching, with much dodging, was required to make our way through them. Here, too, the night soil of the city was being removed in closed receptacles on the shoulders of men, on the backs of horses and oxen and on carts drawn by either. Men and women were hurrying along with baskets of vegetables as illustrated in Fig. 18, some with fresh cabbage, some with high stacks of crisp lettuce, some with monstrous white radishes or turnips, and others with bundles of onions, all coming down from the terraced gardens to the markets. We passed loads of green bamboo poles just cut, 3 inches in diameter at the butt and 20 feet long, drawn on carts. Both men and women were carrying young children, and older children were playing and singing in the street. Many old women, some feeble looking, moved, loaded, through the throng, while homely little dogs, an occasional lean cat, and hens and roosters scurried across the street from one shop to another.

We finally reached one of the terraced hillsides which rise 500 to 1,000 feet above the harbour with sides so steep that gardens have a width of seldom more than 20 to 30 feet and often less, while the front of each terrace is generally a stone wall, sometimes 12, often more than 6, but most commonly 4 and 5 feet high. One of these hillside slopes is seen in Fig. 19. The terraced gardens are both short and narrow and most of them bounded by stone walls on three sides, the two end walls sloping down the hill so as to form footpaths with occasional steps.

Each terrace sloped slightly down the hill and had a low ridge along the front. Around its entire border a narrow drain or furrow was arranged to collect surface water and direct it into

FIG. 18. – Vegetable vendor with his load as carried from house to house.

FIG. 19. – Terraced gardens on hillside at Nagasaki, Japan.

drainage channels or into a catch basin where it might be put back on the garden or be used in preparing liquid fertilizer. At one corner of many of these small terraced gardens were cement-lined pits, used as catch basins for water, as receptacles for liquid manure, or as places in which to prepare compost. Far up the steep paths, along either side, we saw many piles of stable manure awaiting application, all of which had been brought up the slopes in baskets on bamboo poles, carried on the shoulders of men and women.

II

GRAVE LANDS OF CHINA

THE launch had returned the passengers to the steamer at 11.30; the captain was on the bridge; prompt to the minute, at the call 'Hoist away' the signal went below and the *Yamaguchi's* whistle filled the harbour and overflowed the hills. At 12 noon we were leaving Nagasaki, now a city of 153,000 and the western doorway of a nation of 51 millions of people, but of little importance before the sixteenth century, when it became the chief mart of Portuguese trade. We were to pass the Koreans on our right and enter the portals of a third nation of 400 millions. We had left a country which had added 85 millions to its population in one hundred years and which still has twenty acres for each man, woman and child, to pass through one which has but one and a half acres *per capita*, and were going to another whose allotment of acres, good and bad, is less than 2·4. We had come from a country in which three generations had exhausted strong virgin fields, and were approaching others still fertile after thirty centuries of cropping. On January 30th we crossed the head waters of the Mississippi-Missouri, 4,000 miles from its mouth; and on March 1st were in the mouth of the Yangtse River, whose waters are gathered from a basin in which dwell 200 millions of people.

The *Yamaguchi* reached Woosung in the night and anchored to await morning and tide before ascending the Hwangpoo, believed by some geographers to be the middle of three earlier delta arms of the Yangtse-kiang, the southern arm entering the sea at Hangchow 120 miles further south, the third being the present stream. As we wound through this great delta plain toward Shanghai, the city of foreign concessions to all nationalities, the first striking feature was the 'graves of the fathers,' of 'the ancestors.' At first the numerous grass-covered hillocks dotting the plain seemed to be stacks of grain or straw; then came the query whether they might not be huge compost heaps awaiting distribution in the fields, but as the river brought us nearer to them we seemed to be moving through a land of ancient mound builders. Fig. 20 shows, in its upper section, their appearance as seen in the distance.

As the journey led on among the fields, so large were the mounds, often 10 to 12 feet high and 20 or more feet at the base; so grass-covered and apparently neglected; so numerous and so irregularly scattered, without apparent regard for fields, that when we were told that they were graves we could not give credence to the statement. Before the city was reached, however, we saw places where, by the shifting of the channel,

Fig. 20. – Views of grave lands in the delta region of the Yangtse-kiang, China.

the river had cut into some of these mounds, exposing brick vaults. Some were so low as to be under water part of the time, and we wonder if the fact does not also record a slow subsidence of the delta plain under the ever-increasing load of river silt.

A closer view of the graves is given in the lower section of Fig. 20, where they are seen not only to occupy large areas of valuable land but to be much in the way of agricultural operations. A still closer view of other groups, with a farm village in the background, is shown in the middle section of the same illustration. On the right may be seen a line of six graves surmounting a common lower base which is a type of the larger and higher ones so suggestive of buildings seen in the horizon of the upper section.

Everywhere we went in China, especially in the neighbourhood of old and large cities, the proportion of grave land to cultivated fields is very large. In the vicinity of Canton Christian College, on Honam island, more than 50 per cent of the land was given over to graves, and in many places they were so close

FIG. 21. – Goats pasturing on grave lands near Shanghai, and graves in hilly lands near Canton.

that one could step from one to another. They are on the higher and drier lands, the cultivated areas occupying ravines and the lower levels to which water may be more easily applied and which are the most productive. Hilly lands not so readily cultivated, and especially if within reach of cities, are largely so used. These grave lands are not altogether unproductive, for they are generally overgrown with herbage of one kind or another and used as pastures for geese, sheep, goats and cattle, and it is not at all uncommon, when riding along a canal, to see a huge water buffalo projected against the sky from the summit of one of the largest and highest grave mounds within reach. If the herbage is not fed off by animals it is usually cut for feed, for fuel, for green manure or for use in the production of compost to enrich the soil.

FIG. 22. – Cluster of graves in brick vaults, lower section; and isolated grave in garden, with two large grave mounds, upper section.

Caskets may be placed directly upon the surface of a field, encased in brick vaults with tile roofs, forming such clusters as was seen on the bank of the Grand Canal in Chekiang province, represented in the lower section of Fig. 22, or they may stand singly in the midst of a garden, as in the upper section of the same figure; in a paddy field entirely surrounded by water parts of the year, and indeed in almost any unexpected place. In Shanghai in 1898, 2,763 exposed coffined corpses were removed outside the International Settlement or buried by the authorities.

Further north, in the Shantung province, where the dry season is more prolonged and where a severe drought had made grass short, the grave lands had become nearly naked soil, as seen in Fig. 23, where a Shantung farmer had just dug a temporary well to irrigate his little field of barley. Within the range of the camera, as held to take this view, more than forty grave mounds besides the seven near by, are near enough to be fixed on the negative and be discernible under a glass, indicating what extensive areas of land, in the aggregate, are given over to graves.

Still further north, in Chihli, a like story is told in, if possible, more emphatic manner and fully vouched for in the next illustration, Fig. 24, which shows a typical family group, to be observed in so many places between Taku and Tientsin and beyond toward Peking. As we entered the mouth of the Pei-ho for Tientsin, far away to the vanishing horizon there stretched an almost naked plain except for the vast numbers of these 'graves of the fathers,' so strange, so naked, so regular in form and so numerous that more than an hour of our journey had passed before we realized that they were graves and that the country here was perhaps more densely peopled with the dead than with the living. In so many places there was the huge father grave, often capped with what in the distance suggested a chimney, and the many associated smaller ones, that it was difficult to realize in passing what they were.

It is a common custom, even if the residence has been permanently changed to some distant province, to take the bodies back for interment with the family group; and it is this custom which leads to the practice of choosing a temporary location for the body, waiting for a favourable opportunity to remove it. This is the reason often for the isolated coffin so frequently seen under a simple thatch of rice straw.

It is the custom periodically to restore the mounds, maintaining their height and size, as is seen in the next two illustrations, and to decorate them once in the year with flying streamers of coloured

Fig. 23. – Graves surrounded by fields in the Shantung province. The farmer has dug a temporary well to irrigate the little barley field threatened by drought.

Fig. 24. – Family group of grave mounds in Chihli, between Taku and Tientsin; the largest or father grave is in the rear, those of his two sons standing next.

paper, the remnants of which may be seen in both Figs. 25 and 26, set there as tokens that the paper money has been burned upon them and its essence sent up in the smoke for the maintenance of the spirits of the departed. We have our memorial day; they have for centuries observed theirs with religious fidelity.

The usual expense of a burial among the working people is said to be $100, Mexican, an enormous burden when the day's wage or the yearly earning of the family is considered together with the yearly expense of the ancestor worship. How such voluntary burdens are assumed by people under such circumstances is hard to understand. Missionaries assert it is due to fear of evil consequences in this life and of punishment and neglect in the hereafter. Is it not far more likely that such is the price these people are willing to pay for a good name among the living and because of their deep and lasting affection for the departed? Nor does it seem at all strange that a kindly, warm-hearted people with strong filial reverence should have reached, early in their long history, a belief in one spirit of the departed which hovers about the home, one which hovers about the grave, and a third which wanders abroad, for surely there are associations with each of these conditions which must long and forcefully awaken memories of those who are gone. If this view is possible, may not such ancestral worship be an index of qualities of character strongly fixed and of the highest worth which, when improvements come that may relieve the heavy burdens now carried, will only shine more brightly and count more for right living as well as for comfort?

Even in our own case it will hardly be maintained that our burial customs have reached their best and final solution, for in all civilized nations they are unnecessarily expensive and far too cumbersome. It is only necessary mentally to add the accumulation of a few centuries to our cemeteries to realize how impossible our practice must become. Clearly there is here a very important need for betterment which all nationalities should seek.

When the steamer anchored at Shanghai the day was pleasant and the raincoats which greeted us in Yokohama were not in evidence, but the numbers who had met the steamer in the hope of an opportunity for earning a trifle were far greater and in many ways in strong contrast with the Japanese. We were surprised to

Fig. 25. – Grave mounds recently restored and bearing the streamer standards in token of memorial services.

Fig. 26. – Group of grass-grown grave mounds carrying the streamer standards and showing the extensive occupation of land.

find the men of so large a stature, much above the Chinese usually seen in the United States. They were fully the equal of large Americans in frame, but without surplus flesh, though few appeared underfed. To realize that these are strong, hardy men it was only necessary to watch them in pairs carrying on their

Fig. 27. – Men freighters going inland with loads of matches.

shoulders bales of cotton suspended from strong bamboo poles; while the heavy loads they transport on wheelbarrows through the country over long distances, as seen in Fig. 27, prove their great endurance. This same type of vehicle is one of the common means of transporting people, especially women, and four, six and even eight may be seen riding together, propelled by a single wheelbarrow man.

III

TO HONG-KONG AND CANTON

We had come to these countries to learn how the old-world farmers had been able to provide materials for food and clothing on such small areas for so many millions, at so low a price, during so many centuries, and were anxious to see them on the soil and among the crops. The sun was still south of the equator, coming north only about twelve miles per day, so, to save time, we booked on the next steamer for Hong-Kong to meet spring at Canton, beyond the Tropic of Cancer, 600 miles farther south, and return with her.

On the morning of March 4th the *Tosa Maru* steamed out into the Yangtse River, already flowing with the increased speed of ebb tide. The pilots were on the bridge to guide her course along the narrow south channel through waters seemingly as brown and turbid as the Potomac after a rain. It was at some distance beyond Gutzlaff Island that we crossed the front of the out-going tide, showing in a sharp line of contrast across the course of the ship. During long ages this stream of mighty volume has been loading itself in far-away Tibet, without dredge, barge, fuel or human effort, with unused, and there unusable, soils, bringing them down from inaccessible heights a distance of 2,000 or 3,000 miles, building up with them, from the bottom of the sea, at the gateways of commerce, miles upon miles of the world's most fertile fields and gardens. To-day on this river, winding through 600 miles of the most highly cultivated fields, laid out on river-built plains, go large ocean steamers to the city of Hankow-Wuchang-Hanyang, where 1,770,000 people live and trade within a radius of less than four miles; while smaller steamers push on 1,000 miles beyond and are then not more than 130 feet above sea level.

Even now, with the aid of current, tide and man, these brown turbid waters are rapidly adding fertile delta plains for new homes. During the last twenty-five years Chungming Island has grown in length some 1,800 feet per year and to-day 1,000,000 people are living and growing rice, wheat, cotton and sweet potatoes on 270 square miles of fertile plain where 500 years ago were only submerged river sands and silt. Here 3,700 people per square mile have acquired homes.

Sunday morning, March 7th, found us entering the long, narrow and beautiful harbour of Hong-Kong. Here, lying at anchor in the 10 square miles of water, were five battleships, several large ocean steamers, many coastwise vessels and a multitude of smaller craft whose yearly tonnage is twenty to thirty millions. But the harbour lies in the track of the terrible East Indian typhoon and, although sheltered on the north shore of a high island, one of these storms recently sunk nine vessels, sent twenty-three ashore, seriously damaged twenty-one others, wrought great destruction among the smaller craft and destroyed over 1,000 lives. Such was the destruction wrought by the September storm of 1906.

Our steamer did not go to dock, but the Nippon Yusen Kaisha's launch transferred us to a city much resembling Seattle in possessing a scant footing between a long sea front and high steep mountain slopes behind. Here cliffs too steep to climb rise from the very sidewalk and are covered with a profusion of ferns, small bamboo, palms, vines, flowering shrubs, all interspersed with pine and great banyan trees that do so much toward adding the beauty of northern landscapes to that of tropical regions.

Hong-Kong Island is some 11 miles long and 2 to 5 miles wide, while the peak carrying the signal staff rises 1,825 feet above the streets. The Peak tramway, on which, hanging from opposite ends of a strong cable, one car rises up the slope and another descends every fifteen to twenty minutes, affords communication between business houses below and homes in beautiful surroundings and a tempered climate above. Extending along the slopes of the mountains, above the city, are excellent roads, carefully graded, provided with concrete gutters and bridges, along which one may travel on foot, on horseback, by ricksha or sedan chair. Over one of these we ascended along one side of Happy Valley, around its head and down the other side. Only occasionally could we catch glimpses of the summit through the lifting fog, but looking down, across the city and beyond the harbour with its shipping, and up and down the many ravines, the views are among the choicest ever made accessible to the residents of any city. It was the beginning of the migratory season for birds, and trees and shrubbery thronged with many species.

Women were seen engaged in heavy manual labour with the men, carrying crushed rock and sand, for concrete and macadam work, up the steep street slopes long distances from the dock. Like the men, they were of smaller stature than those seen at Shanghai and closely resemble the Chinese in the United States.

Fig. 28. – Usual method of sawing lumber in China.

Both sexes are agile, wiry and strong. Here we first saw lumber-sawing in the open streets after the manner shown in Fig. 28, where wide boards were being cut from camphor logs. In the damp, already warm weather the men were stripped to the waist, their limbs bare to above the knee, and each carried a large towel for wiping away the profuse perspiration.

It was here, too, that we first met the remarkable staging for the erection of buildings of four and six stories, set up without saw, hammer or nail; without injury to or waste of lumber and with the minimum of labour in construction and removal. Poles and bamboo stems were lashed together with overlapping ends, permitting any interval or height to be secured without cutting or nailing, and admitting of ready removal with absolutely no waste,

Fig. 29. – Statuary floral pieces in florist's garden, Happy Valley, Hong-Kong, China.

all parts being capable of repeated use unless it be some of the materials employed in tying members. Up inclined stairways, from staging to staging, in the erection of six-story granite buildings, mortar was being carried in baskets swinging from bamboo poles on the shoulders of men and women, as the cheapest hoists available.

Industrial China, Korea and Japan do not observe our weekly day of rest, and during our walk around Happy Valley on Sunday afternoon, looking down upon its terraced gardens and tiny fields, we saw men and women busy fitting the soil for new crops, gathering vegetables for market, feeding plants with liquid manure and even irrigating certain crops, notwithstanding the damp, foggy, showery weather. Turning the head of the valley, attention was drawn to a walled enclosure, and a detour down the slope brought us to a florist's garden within which were rows of large potted foliage plants of semi-shrubbery habit, seen in Fig. 29, trained in the form of life-size human figures with limbs, arms and trunk provided with highly glazed and coloured porcelain feet, hands and head. These, with many other potted plants and trees, including dwarf varieties, are grown under out-door lattice shelters in different parts of China, for sale to the wealthy Chinese families.

How thorough is the tillage, how efficient and painstaking the garden fitting, and how closely the ground is crowded to its utmost limit of producing power are indicated in Fig. 30; and when one stops and studies the detail in such gardens he expects in its executor an orderly, careful, frugal and industrious man, deriving not a little satisfaction from his creations however arduous his task or prolonged his day. Many were the times, during our walks in the fields and gardens among these old, much misunderstood, misrepresented and undervalued people, when the bond of common interest was felt between us, and we recognized the representative of a race which, with fortitude and rare wisdom, has kept alive the seeds of manhood and nourished them into such sturdy stock.

Not only are these people extremely careful and painstaking in fitting their fields and gardens to receive the crop, but they are even more scrupulous in their care to make everything that can possibly do so serve as fertilizer for the soil, or food for the crop. Expense is incurred to provide such receptacles as are seen in Fig. 31 for receiving not only the night soil of the home and that which may be bought or otherwise procured, but in which may be stored any fluid which can serve as plant food. On the right of these earthenware jars is a pile of ashes and another of manure. All such materials are saved and used in the most advantageous ways to enrich the soil or to nourish the plants in their growth.

Generally the liquid manures must be diluted with water to a greater or less extent before they are 'fed,' as the Chinese say, to

Fig. 30. – A fair type of garden culture seen in Happy Valley, Hong Kong.

their plants, hence there is need of an abundant and convenient water supply. One of these is seen in Fig. 32, where the Chinaman has adopted the modern galvanized iron pipe to bring water from the mountain slope of Happy Valley to his garden. By the side of this tank are the covered pails in which the night soil was brought, perhaps more than a mile, to be first diluted and then applied. But the more general method for supplying water is that

FIG. 31. – Receptacles for collecting liquid manure, and at their right a pile of ashes and a pile of stable manure for fertilizing the garden.

of leading it along the ground in channels or ditches to a small reservoir in one corner of a terraced field or garden, as seen in Fig. 33, where it is held and the surplus led down from terrace to terrace, giving each its permanent supply. At the upper right corner of the engraving may be seen two manure receptacles and a third near the reservoir. The plants on the lower terrace are watercress and those above the same. At this time of the year, on the terraced gardens of Happy Valley, this is one of the crops most extensively grown.

Fig. 32. – Water brought from mountain side in three-quarter inch iron pipe for use in diluting liquid manure and in garden irrigation.

Fig. 33. – Series of terraces, showing small water reservoir near centre, three receptacles for liquid manure, and a bed of watercress in the foreground.

Walking among these gardens and isolated homes, we passed a pig pen provided with a smooth, well-laid stone floor that had just been washed scrupulously clean, like the floor of a house. While I was not able to learn other facts regarding this case, I have little doubt that the washings from this floor had been carefully collected and taken to some receptacle to serve as a plant food.

Looking back as we left Hong-Kong for Canton on the cloudy evening of March 8th, the view was wonderfully beautiful. We were drawing away from three cities, one, electric-lighted Hong-Kong rising up the steep slopes, suggesting a section of sky set with a vast array of stars of all magnitudes up to triple Jupiters; another, old and new Kowloon on the opposite side of the harbour; and between these two, separated from either shore by wide reaches of wholly unoccupied water, lay the third, a mid-strait city of sampans, junks and coastwise craft of many kinds segregated, in obedience to police regulation, into blocks and streets with each setting sun, but only to scatter again with the coming morn. At night, after a fixed hour, no one is permitted to leave shore and cross the vacant water strip except from certain piers and with the permission of the police, who take the number of the sampan and the names of its occupants. Over the harbour three large searchlights were sweeping and it was curious to see the junks and other craft suddenly burst into full blazes of light, like so many monstrous fireflies, to disappear and reappear as the lights came and went. These measures have been taken to lessen the number of cases of foul play in which people have left the wharves at night for some vessel in the strait, never to be heard of again.

The distance by water to Canton is some 90 miles, and early in the morning our steamer dropped anchor off the foreign settlement of Shameen. Through the kindness of Consul-General Amos P. Wilder in sending a telegram to the Canton Christian College, their little steam-launch met the boat and took us directly to the home of the college on Honam Island. The college lies in the great delta south of the city where sediments brought by the Si-kiang, Pei-kiang, and Tung-kiang – west, north and east rivers – through long centuries have been building up the richest of land. This reclaimed land is appropriated as fast as it is formed, and made to bring forth materials for food, fuel and raiment in great quantities.

It was on Honam Island that we walked first among the grave lands and came to know them as such, for Canton Christian College stands in the midst of graves which, although very old, are not permitted to be disturbed. Cattle were grazing and with them some 250 of the brown Chinese geese, two-thirds grown, were gleaning their entire living from the grave lands and adjacent water. A mature goose sells in Canton for $1·20, Mexican, or less

Fig. 34. – Looking across fields which have borne two crops of rice, now ridged for leeks and other vegetables as a winter crop.

than 52 cents, gold, but even then how can the labourer whose day's wage is but 10 or 15 cents afford one for his family?

Here, too, we saw the Chinese persistent, never-ending industry in keeping their land, their sunshine, their rain, and themselves too, busy in producing something needful. Fields which had matured two crops of rice during the long summer, had been laboriously, and largely by hand labour, thrown into strong ridges as seen in Fig. 34, to permit still a third winter crop of some vegetable to be taken from the land.

But this intensive, continuous cropping of the land spells soil exhaustion and creates demands for maintenance and restoration of available plant food or the adding of large quantities of something quickly convertible into it. Here, therefore, in the fields on Honam Island, as in the Happy Valley, there was abundant evidence of the most careful attention and laborious effort devoted to plant feeding. The boat standing in the canal in Fig. 35 had come from Canton in the early morning with two tons of human manure and men were busy applying it, in diluted form,

Fig. 35. – Boat load of human waste in canal on Honam Island, brought from Canton and being used in feeding winter vegetables.

to beds of leeks at the rate of 16,000 gallons per acre, all carried on the shoulders in pairs of pails like those which stand in the foreground. The material is applied with long-handled dippers holding a gallon, the men wading, with bare feet and trousers rolled up above the knees, in the water of the furrows between the beds.

The above is one of the ways of 'feeding the crops' adopted by these agriculturists. They have other methods of 'manuring the soil.' One of these we first met on Honam Island. Large amounts of canal mud are here collected in boats, brought to the fields to be treated and there left to drain and dry before distributing.

Both the material used to feed the crop and that used for manuring the land are waste products, hindrances to the industry of the region, but the Chinese make them do essential duty in maintaining its life. Human waste must be disposed of. We turn it into the sea. They return it to the soil. Doing so, they save for plant feeding more than a ton of phosphorus (2,712 pounds) and more than two tons of potassium (4,488 pounds) per day for each million of adult population. The mud collects in their canals and obstructs movement. They must be kept open. The mud is highly charged with organic matter and adds humus to the soil when applied to the fields, at the same time raising their level above the river and canal, and giving them better drainage. In this way they turn to use what would otherwise be waste, and cause the labour which would be expended in disposal to count in a remunerative way.

During the early morning ride to Canton Christian College and three other rides which we were permitted to enjoy in the launch on the canal and river waters, everything was strange, fascinating and full of human interest. The Cantonese water population was a surprise, not so much for its numbers as for the lithe, sinewy forms, bright eyes and cheerful faces, particularly among the women, young and old. Nearly always one or more women, mother and daughter oftenest, grandmother many times, wrinkled, sometimes grey, but strong, quick and vigorous in motion, were manning the oars of junks, houseboats and sampans. Sometimes husband and wife and many times the whole family were seen together when the craft was both home and business boat as well. Little children were gazing from most unexpected peep-holes, or they toddled tethered from a waist-belt at the end of as much rope as would arrest them above water, should they go overboard. The cat, too, was similarly tied. Through an overhanging latticed stern, hens craned their necks, longing for scenes they could not reach. With bare heads, bare feet, in short trousers and all dressed much alike, men, women, boys and girls showed equal mastery of the oar. Beginning so young, day and night in the open air on the tide-swept streams and canals, exposed to all the sunshine the fogs and clouds permit, and removed from the dust and filth of streets, it would seem that if the children survive at all they must develop strong physiques. The appearance of the women somehow conveyed the impression that they were more vigorous and in better fettle than the men.

Boats selling many kinds of steaming hot dishes were common. Among these were packets of rice tied in green leaf wrappers, in clusters of three suspended by a strand of some vegetable fibre, to be handed hot from the cooker to the purchaser on a passing junk. Another purchaser would buy hot water for a brew of tea, while to another, and for a single cash, would be handed a small square of cotton cloth, wrung out of hot water, with which he would wipe his face and hands and then return it.

Perhaps nothing better measures the intensity of the maintenance struggle here, and indicates the minute economies practised, than the value of their smallest currency unit, the cash, used in their daily retail transactions. On the Pacific coast, where less thought is given to little economies than perhaps anywhere else in the world, the nickel is the smallest coin in general use, twenty of which are equal to the dollar. For the rest of the United States and in most English-speaking countries one hundred cents or halfpennies measure an equal value. In Russia 170 kopecks, in Mexico 200 centavos, in France 250 two-centime pieces, and in Austria-Hungary 250 two-heller coins equal the United States dollar; while in Germany 400 pfennigs, and in India 400 pie are required for an equal value. Again 500 penni in Finland and of stotinki in Bulgaria, of centesimi in Italy and of half cents in Holland, equal the United States dollar. But in China the small daily financial transactions are measured against a much smaller unit, their cash, 1,500 to 2,000 of which are equal to the United States dollar.

In the Shantung province, when we inquired of the farmers the selling prices of their crops, their replies were given like this: 'Thirty-five strings of cash for 420 catties of wheat and twelve to fourteen strings of cash for 1,000 catties of wheat straw.' At this time, according to my interpreter, the value of one string of cash was 40 cents Mexican, from which it appears that something like 250 of these coins were threaded on a string. Twice we saw a wheelbarrow heavily loaded with strings of cash being transported through the streets, lying exposed on the frame, suggesting chains of copper rather than money. At one of the go-downs or warehouses in Tsingtao, where freight was being transferred from a steamer, the carriers were receiving their pay in these coins. The paymaster stood in the doorway with half a bushel of loose cash in a grain sack at his feet. With one hand he received the bamboo tally-sticks from the stevedores and with the other paid out the cash for the service rendered.

Reference has been made to buying hot water. In a sampan managed by a woman and her daughter, who took us ashore, the middle section of the boat was furnished in the manner of a tiny sitting-room, and on the sideboard were teapots in padded baskets suggestive of our fireless cookers, keeping boiled water hot for making tea. This device is centuries old, and boiled water, as tea, is the universal drink, adopted no doubt as a preventive measure against typhoid fever and allied diseases. Few vegetables are eaten raw and nearly all foods are taken hot or recently cooked if not in some way pickled or salted. Houseboat meat shops moved among the many junks on the canals. These were provided with a receptacle, communicating freely with the canal water, in which fish were kept alive until sold. At the street markets too, fish are kept alive in large tubs of water systematically aerated by water falling from an elevated receptacle in a thin stream. Poultry is largely retailed alive although we saw much of it dressed, salted and cooked to a uniform rich brown, hanging exposed in the markets. From facts like these it may be seen that among these people many fundamental sanitary practices are rigidly observed.

The mechanical appliances in use on the canals and in the shops of Canton demonstrate that the Chinese possess constructive ability of a high order, notwithstanding that many of these appliances are of the simplest kind. This statement is well illustrated in the simple yet efficient foot-power shown in Fig. 36, where a father and his two sons are seen to be driving an irrigation pump, lifting water at the rate of seven and a half acre-inches per ten hours, and at a cost, including wage and food, of 36 to 45 cents, gold. On the canals were large stern-wheel passenger boats, capable of carrying from 30 to 100 people, propelled by the same foot-power, the men working in long single or double lines, depending on the size of the boat. On these the fare was one cent, gold, for a 15-mile journey, a rate one-thirtieth of our two-cent railway tariff. The dredging and clearing of the canals and water channels in and about Canton is likewise accomplished by the same foot-power, often by families living on the dredge boats. A dipper dredge is used, constructed of strong bamboo strips woven into the form of a sliding, two-horse road scraper, guided by a long bamboo handle. The dredge is drawn along the bottom by a rope winding

FIG. 36. – The wooden foot-power of China, used to propel the wooden-chain irrigation pump.

about the projecting axle of the foot-power, propelled by three or more people. When the dipper reaches the axle and is raised from the water it is swung aboard, emptied and returned by means of a long arm like the old well sweep, operated by a cord depending from the lower end of the lever, the dipper swinging from the other. Much of the mud so collected from the canals and channels of the city is taken to the rice and mulberry fields, many square miles of which occupy the surrounding country. Thus the channels are kept open, the fields grow steadily higher above flood level, while their productive power is maintained by the plant food and organic matter carried in the sediment.

The mechanical principle involved in the boy's button buzz was applied in Canton and in many other places for operating small drills as well as in grinding and polishing appliances used in the manufacture of ornamental ware. The drill, as used for boring metal, is set in a straight shaft, often of bamboo, on the upper end of which is mounted a circular weight. The drill is driven by a pair of strings with one end attached just beneath the momentum weight and the other fastened at the ends of a cross hand-bar, having a hole at its centre through which the shaft carrying the drill passes. Holding the drill in position for work and turning the shaft, the two cords are wrapped about it in such a manner that simple downward pressure on the hand-bar held in the two hands unwinds the cords and thus revolves the drill. Relieving the pressure at the proper time permits the momentum of the revolving weight to rewind the cords and the next downward pressure brings the drill again into service.

IV

UP THE SI-KIANG, WEST RIVER

On the morning of March 10th we took passage on the *Nanning* for Wuchow, in Kwangsi province, a journey of 220 miles up the West River, or Si-kiang. The *Nanning* is one of two English steamers making regular trips between the two places, and it was the sister boat which in the summer of 1906 was attacked by pirates on one of her trips and all of the officers and first-class passengers killed while at dinner. The cause of this attack, it is said, or the excuse for it, was threatened famine resulting from destructive floods which had ruined the rice and mulberry crops of the great delta region and had prevented the carrying of manure and bean cake as fertilizers to the tea fields in the hill lands beyond, thus bringing ruin to three of the great staple crops of the region. To avoid the recurrence of such tragedies the first-class quarters on the *Nanning* had been separated from the rest of the ship by heavy iron gratings thrown across the decks and over the hatchways. Armed guards stood at the locked gateways, and swords were hanging from posts under the awnings of the first cabin quarters. Both British and Chinese gunboats were patrolling the river; all Chinese passengers, even the Government soldiers, were searched for concealed weapons as they came aboard, and all arms taken into custody until the end of the journey. Several of the large Chinese merchant junks which we passed were armed with small cannon, and when riding by rail from Canton to Sam Shui, a government pirate detective was in our coach.

The Si-kiang is one of the great rivers of China and indeed of the world. Its width at Wuchow at low water was nearly a mile, and our steamer anchored in 24 feet of water to a floating dock made fast by huge iron chains reaching 300 feet up the slope to the city proper, thus providing for a rise of 26 feet in the river at its flood stage during the rainy season. In a narrow section of the river where it winds through Shui Hing gorge, the water at low stage has a depth of more than 25 fathoms, too deep for anchorage, so in times of prospective fog, boats wait for clear weather. On account of these fluctuations in the height of the river vessels passing up to Wuchow are limited to those drawing 6½ feet of water during the low stage, and at high stage to those drawing 16 feet.

When the West River emerges from the high lands, with its burden of silt, to join its waters with those of the North and East rivers, it has entered a vast delta plain some 80 miles from east to west and nearly as many from north to south. This plain has been canalized, diked, drained and converted into the most productive of fields, bearing three or more crops each year. As we passed westward through the delta region the broad flat fields, surrounded by dikes to protect them against high water, were being ploughed and fitted for the coming crop of rice. In many places the dikes were planted with bananas and in the distance gave the appearance of extensive orchards.

At times we approached near enough to the fields to see how they were laid out. From the gates long canals, 6 to 8 feet wide, led back sometimes 80 or 100 rods. Across these and at right angles, head channels were cut and between them the fields were ploughed in long straight lands some 2 rods wide, separated by water furrows. Many of the fields were bearing sugar-cane standing 8 feet high. The Chinese do no sugar refining but boil the sap until it is ready to solidify, when it is run into cakes resembling chocolate or brown maple sugar. Immense quantities of sugar-cane, too, are exported to the northern provinces, in bundles wrapped up in matting or other cover.

In many places the water-course was too broad to permit detailed study of field conditions and crops, even with a glass. In such sections the recent dikes often had the appearance of being built from limestone blocks, but a closer view showed them constructed from blocks of the river silt cut and laid in walls with slightly sloping faces. In time, however, the blocks weather and the dikes become rounded earthen walls.

We passed two men in a boat, in charge of a huge flock of some hundreds of yellow ducklings. Anchored to the bank was a large houseboat with a stack of rice straw and other things which constituted the floating home of the ducks. Both ducks and geese are reared in this manner in large numbers by the river population. When it is desired to move to another feeding ground a gang plank is put ashore and the flock come on board to remain for the night or to be landed at another place.

About five hours' journey westward in this delta plain, where the fields lie 6 to 10 feet above the present water stage, we reached the mulberry district. Here the plants are cultivated in rows about 4 feet apart, having the habit of small shrubs rather than of trees, and so much resembling cotton that our first impression was that we were in an extensive cotton district. On the lower lying areas. surrounded by dikes, some fields were laid out in the

Fig. 37. – Field of mulberry having the surface covered with fresh earth taken from ditches dividing the land into beds.

manner of the old Italian or English water meadows, with a shallow irrigation furrow along the crest of the bed and deeper drainage ditches along the division line between them. Mulberries were occupying the ground before the freshly cut trenches we saw were dug, and all the surface between the rows had been evenly overlaid with the fresh earth removed with the spade, the soil lying in blocks essentially unbroken. In Fig. 37 may be seen the mulberry crop on a surface treated in this way, between Canton and Sam Shui, with the earth removed from the trenches laid evenly over the entire surface between and around the plants.

At frequent intervals along the river, paths and steps were seen leading to the water and within a distance of a quarter of a mile we counted thirty-one men and women carrying mud in baskets on bamboo poles swung across their shoulders, the mud being taken from just above the water line. The disposition of this material we could not see, as it was carried beyond a rise in ground. We have little doubt that the mulberry fields were being covered with it. It was here that a rain set in and almost like magic the fields blossomed out with great numbers of giant rain hats and kittysols, where people had been unobserved before. From one o'clock until six in the afternoon we had travelled continuously through these mulberry fields stretching back miles from our line of travel on either hand, and the total acreage must have been very large. But we had now nearly reached the margin of the delta and the mulberries were exchanged for fields of grain, beans, peas and vegetables.

After leaving the delta region, the balance of the journey to Wuchow was through hill country, the slopes rising steeply from near the river bank, so that there was relatively little tilled or readily tillable land. Rising usually 500 to 1,000 feet, the sides and summits of the rounded, soil-covered hills were generally clothed with a short herbaceous growth and small scattering trees, oftenest pine, 4 to 16 feet high.

In several sections along the course of the river there are limited areas of intense erosion where naked gulleys of no mean magnitude have developed; but these were exceptions, and we were continually surprised at the remarkable steepness of the slopes, with convexly rounded contours almost everywhere, well mantled with soil, devoid of gulleys and completely covered with herbaceous growth dotted with small trees. The absence of forest growth finds its explanation in human influence rather than natural conditions.

Throughout the hill-land section of this mighty river the most characteristic and persistent human features were the stacks of brushwood and the piles of firewood along the banks or loaded upon boats and barges for the market. The brush-wood was largely made from the boughs of pine, tied into bundles and stacked like grain. The firewood was usually round, peeled and made from the limbs and trunks of trees 2 to 5 inches in diameter.

All this fuel was coming to the river from the back country, sent down along steep slides which in the distance resemble paths leading over hills but too steep for travel. The fuel was loaded upon large barges, the boughs in the form of stacks to shed rain but with a tunnel leading into the house of the boat about which they were stacked, while the wood was similarly corded about the dwelling. The wood was going to Canton and other delta cities, while the pine boughs were taken to the lime and cement kilns, many of which were located along the river. Absolutely the whole tree, including the roots and the needles, is saved and burned; no waste is permitted.

The up-river cargo of the *Nanning* was chiefly matting rush, taken on at Canton, tied in bundles like sheaves of wheat. It is grown upon the lower, newer delta lands by methods of culture similar to those applied to rice. Fig. 38 shows a field as seen in Japan.

The rushes were being taken to one of the country villages on a tributary of the Si-kiang and the steamer was met by a flotilla of junks from this village, some 45 miles up the stream, where the families live who do the weaving. On the return trip the flotilla again met the steamer with a cargo of the woven matting. In keeping record of packages transferred the Chinese use a simple and unique method. Each carrier, with his two bundles, received a pair of tally sticks. At the gang-plank sat a man with a tally-case divided into twenty compartments, each of which could receive five, but no more, tallies. As the bundles left the steamer the tallies were placed in the tally-case until it contained one hundred, when it was exchanged for another.

Wuchow is a city of some 65,000 inhabitants, standing back on the higher ground, not readily visible from the steamer landing nor from the approach on the river. On the foreground, across which stretched the anchor chains of the dock, was living a floating population, in shelters less substantial than Indian wigwams, but engaged in a great variety of work, and many water buffalo had been tied for the night along the anchor chains. Before July much of this area would lie beneath the flood waters of the river.

Here a ship-builder was using his simple effective bow-brace, boring holes for the dowel pins in the planking for his ship, and

FIG. 38. – Landscape in Japan, showing fields of rice and matting rush. The dark area in the foreground and others back are occupied by the rush.

another was bending the plank to the proper curvature. The bow-brace consisted of a bamboo stalk carrying the bit at one end and a shoulder rest at the other. Pressing the bit to its work with the shoulder, it was driven with the string of a long bow wrapped once around the stalk by drawing the bow back and forth, thus rapidly and readily revolving the bit.

The bending of the long, heavy plank, 4 inches thick and 8 inches wide, was more simple still. It was saturated with water

Fig. 39. – Wooden fork shaped from the limbs of a tree by simple means of steaming and drying.

and one end raised on a support 4 feet above the ground. A bundle of burning rice straw moved along the under side against the wet wood had the effect of steaming the wood and the weight of the plank caused it to gradually bend into the shape desired. Bamboo poles are commonly bent or straightened in this manner to suit any need and Fig. 39 shows a wooden fork shaped in the manner described from a small tree having three main branches. This fork is in the hands of my interpreter and was used by the woman standing at the right, in turning wheat.

When the old ship-builder had finished shaping his plank he

sat down on the ground for a smoke. His pipe was one joint of bamboo stem 1 foot long, nearly 2 inches in diameter and open at one end. In the closed end, at one side, a small hole was bored for draught. A charge of tobacco was placed in the bottom, the lips pressed into the open end and the pipe lighted by suction, holding a lighted match at the small opening. To enjoy his pipe the bowl rested on the ground between his legs. With his lips in the bowl and a long breath, he would completely fill his lungs, retaining the smoke for a time, then slowly expire and fill the lungs again, after an interval of natural breathing.

On returning to Canton we went by rail, with an interpreter, to Sam Shui, visiting fields along the way. Fig. 40 is a view of one landscape. The woman was picking roses among tidy beds of garden vegetables. Beyond her and in front of the near building are two rows of waste receptacles. In the centre background is a large 'go-down,' in function that of our cold storage warehouse and in part that of our grain elevator for rice. In them, too, the wealthy store their fur-lined winter garments for safe keeping. These are numerous in this portion of China and the rank of a city is indicated by their number. The conical hillock is a large near-by grave mound and many others serrate the sky-line on the hill beyond.

In the next landscape, Fig. 41, a crop of winter peas, trained to canes, are growing on ridges among the stubble of the second crop of rice. In front is one canal, the double ridge behind is another and a third canal extends in front of the houses. Already preparations were being made for the first crop of rice, fields were being flooded and fertilized. One such is seen in Fig. 42, where a labourer was engaged at the time in bringing stable manure, wading into the water to empty the baskets.

Two crops of rice are commonly grown each year in southern China, and during the winter and early spring, grain, cabbage, rape, peas, beans, leeks and ginger may occupy the fields as a third or even fourth crop, making the total year's product from the land very large; but the amount of thought, labour and fertilizers given to securing these is even greater. How great these efforts are will be appreciated from what is seen in Fig. 43, representing two fields thrown into high ridges, planted to ginger and covered with straw. All of this work is done by hand and

Fig. 40. – Landscape at Sam Shui, near Canton.

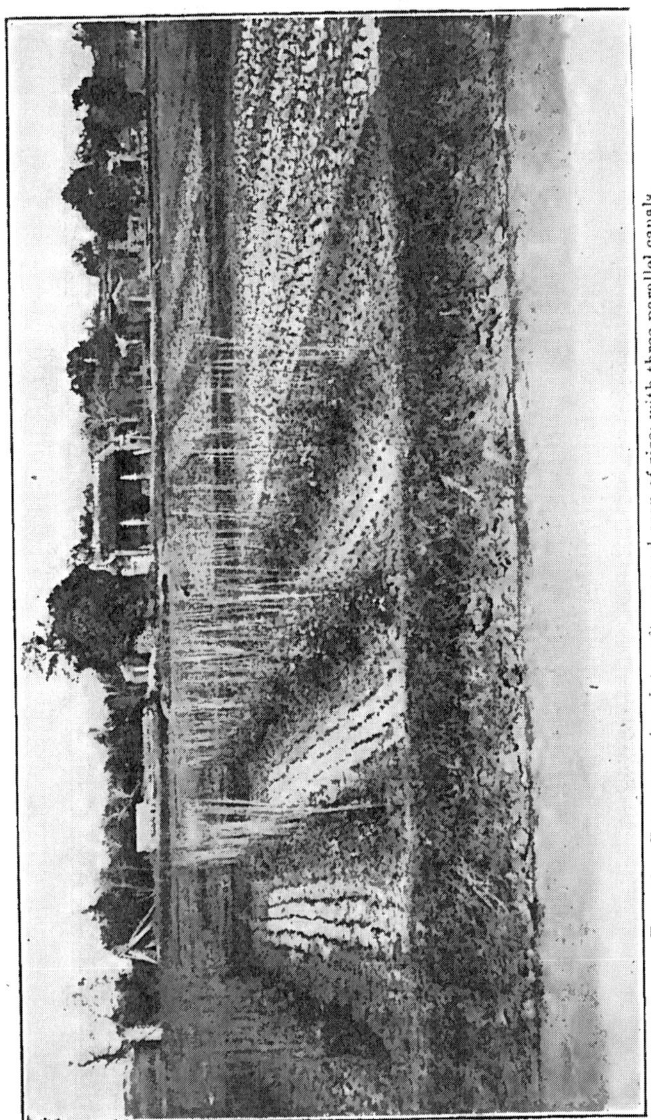

FIG. 41. – Peas grown in winter after second crop of rice; with three parallel canals.

Fig. 42. — Flooded fields being fertilized preparatory for rice; winter peas and beans beyond, with dwellings in the background.

when the time for rice planting comes every ridge will again be thrown down and the surface smoothed to a water level. Even when the ridges and beds are not thrown down for the crops of

Fig. 43. – Fields of ginger just planted; ridged and furrowed for drainage, showing the amount of hand labour performed to secure the winter crop, following two of rice.

rice, the furrows and the beds will change places so that all the soil is worked over deeply and mainly through hand labour. The statement so often made, that these people only barely scratch the surface of their fields with the crudest of tools, is very far from the truth, for their soils are worked deeply and often, notwithstanding the fact that their ploughing, as such, may be shallow.

Through Dr. John Blumann of the missionary hospital at Tungkun, east from Canton, we learned that the good rice lands there a few years ago sold at $75 to $130 per acre, but that prices are rising rapidly. The holdings of the better class of farmers there are 10 to 15 mow – $1\frac{2}{3}$ to $2\frac{1}{2}$ acres – upon which are maintained families numbering six to twelve. The day's wage of a carpenter or mason is 11 to 13 cents, U.S. currency, board not included, but a day's ration for a labouring man is counted worth 15 cents, Mexican, or less than 7 cents, gold.

Fish culture is practised in both deep and shallow basins, the deep permanent ones renting as high as $30, gold, per acre. The shallow basins which can be drained in the dry season are used for fish only during the rainy period, being later drained and planted to some crop. The permanent basins have often come to be 10 or 12 feet deep, increasing with long usage, for they are periodically drained by pumping and the foot or two of mud which has accumulated removed and sold as fertilizer to planters of rice and other crops. It is a common practice, too, among the fish growers, to fertilize the ponds, and in case a footpath leads alongside, screens are built over the water to provide accommodation for travellers. Fish reared in the better fertilized ponds bring a higher price in the market. The fertilizing of the water favours a stronger growth of food forms, both plant and animal, upon which the fish live, and thus they are better nourished, and make a more rapid growth, as is the case with well-fed animals.

V

EXTENT OF CANALIZATION AND SURFACE FITTING OF FIELDS

ON the evening of March 15th we left Canton for Hong-Kong and the following day embarked again on the *Tosa Maru* for Shanghai. Although our steamer was generally out of sight of land except for some off-shore islands, the water was turbid most of the way after we had crossed the Tropic of Cancer off the mouth of the Han River at Swatow. Over a sea bottom measuring more than 600 miles northward along the coast, and perhaps 50 miles to sea, unnumbered acre-feet of the richest soil of China are being borne beyond the reach of her 400 millions of people and the children to follow them. Surely it must be one of the great tasks of future statesmanship and engineering skill to divert larger amounts of such sediments close along inshore in such a manner as to add valuable land annually to the public domain.

In the vast Cantonese delta plains which we had just left, in the still more extensive ones of the Yangtse-kiang to which we were now going, and in those of the shifting Hwang-ho further north, centuries of toiling millions have executed works of almost incalculable magnitude, fundamentally along such lines as those just suggested. They have accomplished an enormous share of these tasks by sheer force of body and will, building levees, digging canals, diverting the turbid waters of streams through them and then carrying the deposits of silt and organic growth out upon the fields, often upon the shoulders of men in the manner we have seen.

It is wellnigh impossible, by word or map, to convey an adequate idea of the magnitude of the systems of canalization and delta and other lowland reclamation work, or of the extent of surface-fitting of fields which have been effected in China, Korea and Japan through many centuries, and which are still in progress. The lands so reclaimed and fitted constitute their most enduring asset and support their densest populations. In one of our journeys by houseboat on the delta canals between Shanghai and Hangchow, over a distance of 117 miles, we made a careful record of the number and dimensions of lateral canals entering and leav-

FIG. 44. – Map of main canals in 718 square miles of Chekiang Province. Each line represents a canal.

ing the main canal along which our boat-train was travelling. This record shows that in 62 miles, beginning north of Kashing and extending south to Hangchow, there entered from the west 134 and there left on the coast side 190 canals. The average width of these canals, measured along the water line, we estimated at 22 and 19 feet on the two sides respectively. The height of the fields above the water level ranged from 4 to 12 feet, during the April and May stage of water. The depth of water, after we entered the

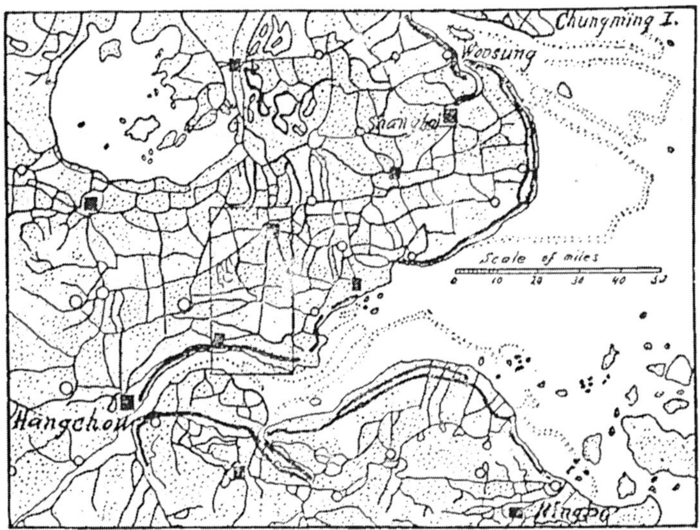

Fig. 45. – Sketch map of portions of Chekiang and Kiangsu Provinces, representing some 2,700 miles of main canals and over 300 miles of sea-wall. The sea-walls are represented by the very heavy black lines. The small rectangle shows the area covered by Fig. 44.

Grand Canal, often exceeded 6 feet, and our best judgment would place the average depth of all canals in this part of China at more than 8 feet below the level of the fields.

In Fig. 44, representing an area of 718 square miles in the region traversed, all the lines are canals, but scarcely more than one-third of the actual number are shown. Between A, where we began our records, before reaching Kashing, and B, near the left margin of the map, there were forty-three canals leading in from the up-country side, instead of the eight shown, and on the coast side there were eighty-six leading water out into the delta plain toward the coast, instead of the twelve shown. Again, on one of our trips by rail, from Shanghai to Nanking, we made a similar record of the number of canals seen from the train, close along the track, and the notes show, in a distance of 162 miles, 593 canals between Lungtan and Nansiang. This is an average of more than three canals per mile for this region and that between Shanghai and Hangchow.

The extent, nature and purpose of these vast systems of internal improvement may be better realized through a study of the next two sketch maps. The first, Fig. 45, represents an area 175 by 160 miles, of which the last illustration is the portion enclosed in the small rectangle. On this area there are shown 2,700 miles of canals, and only about one-third of the canals shown in Fig. 44 are laid down on this map. According to our personal observations there are three times as many canals as are shown on the map of which Fig. 44 represents a part. It is probable, therefore, that there exists to-day in the area of Fig. 45 not less than 25,000 miles of canals.

In the next illustration, Fig. 46, an area of north-east China, 600 by 725 miles, is represented. The unshaded land area covers nearly 200,000 square miles of alluvial plain. This plain is so level that at Ichang, nearly a thousand miles up the Yangtse, the elevation is only 130 feet above the sea. The tide is felt on the river to beyond Wuhu, 375 miles from the coast. During the summer the depth of water in the Yangtse is sufficient to permit ocean vessels drawing 25 feet of water to ascend 600 miles to Hankow, and for smaller steamers to go on to Ichang, 400 miles further.

The location, in this vast low delta and coastal plain, of the system of canals already described, is indicated by the two rectangles in the south-east corner of the sketch map, Fig. 46. The heavy barred black line extending from Hangchow in the south to Tientsin in the north represents the Grand Canal which has a length of more than 800 miles. Westward, up the Yangtse valley, the provinces of Anhwei, Kiangsi, Hunan and Hupeh have very extensive canalized tracts. Still further west, in Szechwan province, is the Chengtu plain, 30 by 70 miles, with what has been called 'the most remarkable irrigation system in China.'

Westward beyond the limits of the sketch map, up the Hwang-ho valley, there is a reach of 125 miles of irrigated lands about

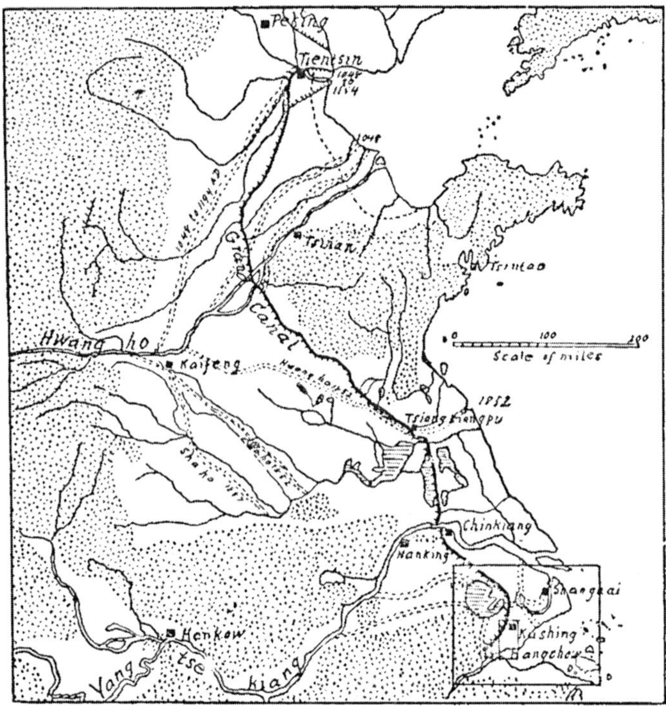

Fig. 46.—Sketch map of north-east China showing the alluvial plain and the Grand Canal, extending 800 miles from Hangchow to Tientsin. The unshaded land area lies mostly less than 100 feet above sea level.

Ninghaifu, and others still farther west, at Lanchowfu and at Suchow where the river has attained an elevation of 5,000 feet, in Kansu province; and there is still to be named the great Canton delta region. A conservative estimate would place the miles of canals and leveed rivers in China, Korea and Japan equal to eight times the number represented in Fig. 45. Fully 200,000 miles in all. Forty canals across the United States from east to west and sixty from north to south would not equal in number of miles those in these three countries to-day. Indeed, it is probable that this estimate is not too large for China alone.

As adjuncts to these vast canalization works there have been enormous amounts of embankment, dike and levee construction. More than 300 miles of sea wall alone exist in the area covered by the sketch map, Fig. 45. The east bank of the Grand Canal, between Yangchow and Hwaianfu, is itself a great levee, holding back the waters to the west above the eastern plain, and diverting them south, into the Yangtse-kiang. But it is also provided with spillways to permit waters to discharge eastward in times of excessive flood. Such excess waters, however, are controlled by another dike with a canal along its west side, some 40 miles to the east, which impounds the water in a series of large lakes until it may gradually drain away. This area is seen in Fig. 46, north of the Yangtse River.

Along the banks of the Yangtse, and for many miles along the Hwang-ho, great levees have been built. Sometimes they are in reinforcing series of two or three at different distances back from the channel where the stream bed is above the adjacent country. Their purpose is to limit the inundated areas in times of unusual flood, and so prevent the disaster from becoming widespread. In the province of Hupeh, where the Han River flows through 200 miles of low country, this stream is diked on both sides throughout the whole distance, and in a portion of its course the height of the levees reaches thirty feet or more. Again, in the Canton delta region there are other hundreds of miles of sea wall and dikes, so that the aggregate mileage of this type of construction works in the Empire can only be measured in thousands of miles.

In addition to the canal and levee construction works there are numerous impounding reservoirs which are brought into requisition to control overflow waters from the great streams. Two of these reservoirs, Tungting lake in Hupeh and Poyang in Hunan, have areas of 2,000 and 1,800 square miles respectively, and during the heaviest rainy seasons each may rise through 20 to 30 feet. Then there are other large and small lakes in the coastal plain giving an aggregate reservoir area exceeding 13,000 square miles. All these are brought into service in controlling flood waters, and are steadily filling with the sediments brought from the far-away uncultivable mountain slopes. They are ultimately destined to become rich alluvial plains, doubtless to be canalized in the manner we have seen.

There is still another phase of these vast construction works which has been of the greatest moment in increasing the maintenance capacity of the Empire – the wresting from the flood waters of the enormous volumes of silt which they carry, depositing it over the flooded areas, in the canals and along the shores in such manner as to add to the habitable and cultivable land. Reference has been made to the rapid growth of Chungming Island in the mouth of the Yangtse-kiang, and the million people now finding homes on the 270 square miles of newly made land which now has its canals, as may be seen in the upper margin of Fig. 45. The city of Shanghai, as its name signifies, stood originally on the seashore, which has now grown 20 miles northward and eastward. In 220 B.C. the town of Putai in Shantung stood one-third of a mile from the sea, but in 1730 it was 47 miles inland, and is 48 miles from the shore to-day.

Sienshuiku, on the Pei-ho, stood upon the seashore in A.D. 500. We passed the city, on our way to Tientsin, 18 miles inland. The dotted line laid in from the coast of the Gulf of Chihli in Fig. 46 marks one historic shore line and indicates a general growth of land 18 miles to seaward.

Besides these actual extensions of the shore lines the centuries of flooding of lakes and low-lying lands has so filled many depressions as to convert large areas of swamp into cultivated fields. Not only this, but the spreading of canal mud broadcast over the encircled fields has had two very important effects – namely, raising the level of the low-lying fields, giving them better drainage and so better physical condition, and adding new plant food in the form of virgin soil of the richest type, thus contributing to the maintenance of soil fertility, high maintenance capacity and permanent agriculture through the centuries.

These operations of maintenance and improvement had a very early inception; they appear to have persisted throughout the recorded history of the Empire and are in vogue to-day. Canals of the type illustrated in Figs. 44 and 45 have been built between 1886 and 1901, both on the extensions of Chungming Island and the newly formed mainland to the north, as is shown by comparison of Stieler's atlas, revised in 1886, with the recent German survey.

Earlier than 2255 B.C., more than 4,100 years ago, Emperor Yao appointed 'The Great' Yu 'Superintendent of Works' and entrusted him with the work of draining off the waters of disastrous floods and of canalizing the rivers. He devoted thirteen years to this work. He was finally called, much against his wishes, to serve as Emperor during the last years of his life.

The history of the Hwang-ho is one of disastrous floods and shiftings of its course, which have occurred many times since the time of the Great Yu, who perhaps began the works perpetuated to-day. Between A.D. 1300 and 1852 the Hwang-ho emptied into the Yellow Sea south of the highlands of Shantung, but in that year, when in unusual flood, it broke through the north levees and finally took its present course, emptying again into the Gulf of Chihli, some 300 miles further north. Some of these shiftings of course of the Hwang-ho and of the Yangtse-kiang are indicated in dotted lines on the sketch map, Fig. 46, where it may be seen that the Hwang-ho during 146 years, poured its waters into the sea as far north as Tientsin, through the mouth of the Pei-ho, 400 miles to the north of its mouth in 1852.

This mighty river is said to carry at low stage, past the city of Tsinan in Shantung, no less than 4,000 cubic yards of water per second, and three times this volume when running at flood. This is water sufficient to inundate 33 square miles of level country 10 feet deep in twenty-four hours. What must be said of the mental status of a people who for forty centuries have measured their strength against such a Titan racing past their homes above the level of their fields, confined only between walls of their own construction? While they have not always succeeded in controlling the river, they have never failed to try again. In 1877 this river broke its banks, inundating a vast area, bringing death to a million people. Again, as late as 1898, 1,500 villages to the north-east of Tsinan and a much larger area to the south-west of the same city were devastated by it, and it is such events as these which have won for the river the names 'China's Sorrow,' 'The Ungovernable' and 'The Scourge of the Sons of Han.'

The construction of the Grand Canal appears to have been a

comparatively recent event in Chinese history. The middle section, between the Yangtse and Tsingkiangpu, is said to have been constructed about the sixth century B.C.; the southern section, between Chingkiang and Hangchow, during the years A.D. 605 to 617 ; but the northern section, from the channel of the Hwang-ho deserted in 1852, to Tientsin, was not built until the years 1280–83.

While this canal, called by the Chinese Yu-ho (Imperial river), Yun-ho (Transport river) or Yunliang-ho (Tribute-bearing river), has connected the great rivers coming down from the far interior into a great water-transport system, it may have been but one of many products of a dominating purpose, namely, the maintenance of the increasing flood of humanity. And I am willing to grant to the Great Yu, with his finger on the pulse of the nation, the power to project his vision 4,000 years into the future of his race and to formulate some of the measures which might be inaugurated through the years and ensure perpetual maintenance for those to follow.

The exhaustion of cultivated fields must always have been the most fundamental and difficult problem of all civilized people, and it appears clear that such canalization as is illustrated in Figs. 44 and 45 may primarily have been initial steps in the reclamation of delta and overflow lands. At any rate, the canalization of the delta and overflow plains of China has been one of the most fundamental and fruitful measures for the conservation of her national resources that they could have taken, and we are convinced that this oldest nation in the world has thus greatly augmented the extension of its coastal plains, conserving hundreds of square miles of the richest and most enduring of soils. We believe, too, that were a full and accurate account given of human influence upon the changes in this remarkable region during the last 4,000 years, it would show that these gigantic systems of canalization have been matters of slow, gradual growth; and that they have been initiated and profoundly influenced by the labours of the strong, patient, persevering, thoughtful but ever silent husbandmen in their efforts to acquire homes and to maintain the productive power of their fields.

Nothing appears clearer than that the greatest material problem which can engage the best thought of China to-day is that of perfecting and extending the means for controlling her flood waters. With her millions of people needing homes and anxious to earn a livelihood the Government should give serious thought to the possibility of putting large numbers of them to work, effectively directed by the best engineering skill. It must now be entirely practicable, with engineering skill and mechanical appliances, to put the Hwang-ho, and other rivers of China subject to overflow, completely under control. With the Hwang-ho confined to its channel, the adjacent low lands can be better drained by canalization and freed from the accumulating saline deposits which are rendering them sterile. Warping might be resorted to during the flood season to raise the level of adjacent low-lying fields, rendering them at the same time more fertile. Where the river is running above the adjacent plains there would be no difficulty in drawing off the turbid water by gravity, under controlled conditions, into diked basins, and even in compelling the river to buttress its own levees. There is certainly great need and great opportunity for China to make still better and more efficient her already wonderful transportation canals and those devoted to drainage, irrigation and fertilization.

In the United States, along the same lines, now that we are considering the development of inland waterways, the subject should be surveyed broadly and much careful study may well be given to the works these ancient peoples in the East have developed and found serviceable through so many centuries. The Mississippi is annually bearing to the sea nearly 225,000 acre-feet of the most fertile sediment, and between levees along a raised bed through 200 miles of country subject to inundation. The time has come when there should be undertaken a systematic diversion of a large part of this fertile soil over the swamp areas, building them into well drained, cultivable, fertile fields provided with waterways to serve for drainage, irrigation, fertilization and transportation. These great areas of swamp land might thus be converted into the most productive rice and sugar plantations to be found anywhere in the world, and the area made capable of maintaining many millions of people as long as the Mississippi endures.

But the conservation and utilization of the wastes of soil erosion, as applied in the delta plain of China, stupendous as this work has been, is nevertheless small when measured by the savings which accrue from the careful and extensive fitting of fields so largely practised, which both lessens soil erosion and permits a large amount of soluble and suspended matter in the run-off to be applied to the fields. Mountainous and hilly as are the lands of Japan, 11,000 square miles of her cultivated fields in the main islands of Honshu, Kyushu and Shikoku have been carefully graded to water level areas, bounded by narrow raised rims, upon which sixteen or more inches of run-off water, with its suspended and soluble matters, may be applied, a large part of which is retained on the fields or utilized by the crop, while surface erosion is almost completely prevented. The illustrations, Figs. 9, 10 and 11, show the application of the principle to the larger and more level fields, and in Figs. 134, 135 and 195 may be seen the practice on steep slopes.

If the total area of fields graded practically to a water level in Japan aggregates 11,000 square miles, the total area thus surface fitted in China must be eight or tenfold this amount. Such enormous field erosion as is tolerated at the present time in our southern and south Atlantic States is permitted nowhere in the Far East, so far as we observed, not even where the topography is much steeper. The tea orchards as we saw them on the steeper slopes, not level-terraced, are often heavily mulched with straw, which makes erosion, even by heavy rains, impossible. This treatment retains the rain where it falls, giving the soil opportunity to receive it under the impulse of both capillarity and gravity, and with it the soluble ash ingredients leached from the straw. The straw mulches we saw used in this manner were often 6 to 8 inches deep, and constituted a dressing of not less than 6 tons per acre, carrying 140 pounds of soluble potassium and 12 pounds of phosphorus. The practice, therefore, gives at once good fertilization, the highest conservation and utilization of rainfall, and a complete protection against soil erosion.

In the Kiangsu and Chekiang provinces, as elsewhere in the densely populated portions of the Far East, we found almost all the cultivated fields nearly level or made so by grading. An instance showing the type of grading in a comparatively level country is seen in Fig 47. By this preliminary surface fitting of the fields the people have reduced to the lowest possible limit the

Fig. 47. – Fields graded for the better conservation of rainfall and fertility and for more efficient irrigation and drainage, Chekiang province.

waste of soil fertility by erosion and surface leaching. At the same time they are able to retain upon the field the largest part of the rainfall practicable, and to compel a much larger proportion of the necessary run-off to leave by under-drainage than would be possible otherwise, thus conveying the plant food developed in the surface soil to the roots of the crops, while they make possible a more complete absorption and retention by the soil of the soluble plant food materials not taken up. This same treatment also furnishes the best possible conditions for the application of water to the fields when supplemental irrigation is helpful, and for the withdrawal of surplus rainfall by surface drainage, when this is necessary.

Besides the surface fitting of fields there is a wide application of additional methods aiming to conserve both rainfall and soil fertility, one of which is illustrated in Fig. 48, showing one end of a collecting reservoir. There were three of these reservoirs, connected with each other by surface ditches and with an adjoining canal. About the reservoir the level field is seen to be thrown into beds with shallow furrows between the long narrow ridges. The furrows are connected by a head drain around the margin of the reservoir and separated from it by a narrow raised rim. Such a reservoir may be 6 to 10 feet deep but can be completely drained only by pumping or by evaporation during the dry season. Into such reservoirs the excess surface water is drained and thus all suspended matter carried from the field is collected and returned, either directly as an application of mud or as material used in composts. In the preparation of composts, pits are dug near the margin of the reservoir, as seen in the illustration, and into them are thrown coarse manure and any roughage in the form of stubble or other refuse which may be available, these materials being saturated with the soft mud dipped from the bottom of the reservoir.

In all the provinces where canals are abundant they also serve as reservoirs for collecting surface washings, and along their banks great numbers of compost pits are maintained and repeatedly filled during the season, for use on the fields as the crops are changed. Fig. 49 shows two such pits on the bank of a canal, already filled.

In other cases, as in the Shantung province, illustrated in Fig. 50,

Fig. 48. – Collecting reservoir for the conservation of rainfall and fertility, used also as fish ponds and to provide water and mud for making composts. The circular ring in the foreground is a compost pit.

Fig. 49. – Two compost pits filled with roughage and mud from the canal, in preparation of compost for the fields. The narrow path along the canal is one of the common thoroughfares in Kiangsu province.

the surface of the field may be thrown into broad levelled lands separated and bounded by deep and wide trenches into which the excess water of very heavy rains may collect. As we

Fig. 50. – Trenching of fields for drainage, conservation of rainfall and of fertility, in the Shantung province. Trenches are 2 feet wide on the bottom, 6 or 8 feet wide at the top, and $2\frac{1}{2}$ to 3 feet deep.

saw them there was no provision for draining the trenches and the water thus collected either seeps away or evaporates, or it may be returned in part by underflow and capillary rise to the soil from which it has been collected. In this province the rains may often be heavy but the total fall for the year is small, being little more than 24 inches, hence there is the greatest need for the conservation so carefully practised.

VI

SOME CUSTOMS OF THE COMMON PEOPLE

The *Tosa Maru* brought us again into Shanghai on March 20th, just in time for the first letters from home. A ricksha man carried us and our heavy valise at a smart trot from the dock to the Astor House, more than a mile, for 8·6 cents, U.S. currency, and more than the conventional price for the service rendered. On our way we passed several loaded barrows, on which women were riding for a fare one-tenth of that which we had paid, but at a slower pace and with many a jolt.

The ringing chorus which came loud and clear when yet half a block away announced that the pile drivers were still at work on the foundation for an annexe to the Astor House. On May 27th, when we returned from the Shantung province, 88 days after we saw them first, they were still at the same work but with the task then practically completed. Had the eighteen men laboured continuously through this interval, the cost of their services to the contractor would have been but $205·92. With these conditions the engine-driven pile driver could not compete. All ordinary labour here receives a low wage. In the Chekiang province farm labour employed by the year received $30, silver, and board, ten years ago, but now is receiving $50. This is at the rate of about $12·90 and $21·50, gold, materially less than is paid per month in the United States. At Tsingtao in the Shantung province a missionary was paying a Chinese cook $10 per month, a man for general work $9 per month, and the cook's wife, for doing the mending and other family service, $2 per month, all living at home and feeding themselves. This service, rendered for $9·03, gold, per month, covers the marketing, all care of the garden and lawn as well as all the work in the house. Missionaries in China find such servants reliable and satisfactory, and trust them with the purse and the marketing for the table, finding them not only honest but far better at a bargain and at economical selection than themselves.

We had a soil tube made in the shops of a large English shipbuilding and repair firm, employing many hundred Chinese as mechanics, using the most modern and complex machinery, and the foreman stated that as soon as the men could understand well enough to take orders they were even better shop hands than the average in Scotland and England. An educated Chinese booking clerk at the Soochow railway station in Kiangsu province was receiving a salary of $10·75, gold, per month. We had inquired the way to the Elizabeth Blake hospital, and he volunteered to escort us and did so, the distance being over a mile. He would accept no compensation, and yet I was an entire stranger, without introduction of any kind.

Everywhere we went in China, the labouring people appeared happy and contented, and showed clearly that they were well nourished. The industrial classes are thoroughly organized, having had their guilds or labour unions for centuries. Nowhere among these densely crowded people, either Chinese, Japanese or Korean, did we see one intoxicated. All classes and both sexes use tobacco, and the British-American Tobacco Company does a business in China amounting to millions of dollars annually.

Among the most frequent sights in the city streets are the itinerant venders of hot foods and confections. Stove, fuel, supplies and appliances may all be carried on the shoulders, swinging from a bamboo pole. The printing of calico by a simple yet effective method handed down through many generations was one of the sights which arrested us. The printer was standing at a rough bench upon which a large heavy stone in cubical form served as a weight to hold in place a thoroughly lacquered sheet of tough cardboard in which was cut the pattern to appear in white on the cloth. Beside the stone stood a pot of thick paste prepared from a mixture of lime and soy bean flour. The soy beans were being ground in a corner of the same room by a diminutive edition of such an outfit as seen in Fig. 51. The donkey was working in his permanent abode and whenever off duty he halted before manger and feed. At the operator's right lay a bolt of white cotton cloth fixed to unroll and pass under the stencil, held stationary by the heavy weight. To print, the stencil was raised and the cloth brought to place under it. The paste was then deftly spread with a paddle over the surface and thus upon the cloth beneath wherever exposed through the openings in the stencil. This completes the printing of the pattern on one section of the bolt of cloth. The free end of the stencil is then raised, the cloth passed along the proper distance by hand and the stencil dropped in place for the next application. The paste is permitted to dry upon the cloth and when the bolt has been dipped into the blue dye the portions protected by the paste remain white. In this simple manner has the printing of calico been done for centuries for the garments of millions of children. From the ceiling of the drying-room were hanging some hundreds of stencils bearing different patterns. In our great calico mills, printing hundreds of yards per minute, the

Fig. 51. – Stone mill in common use for grinding beans and various kinds of grain.

mechanics and the chemistry differ from this primitive method only in detail of application and in dispatch, not in fundamental principle.

In almost any direction we travelled outside the city, in the pleasant mornings when the air was still, the laying of warp for cotton cloth could be seen, to be woven later in the country homes. We saw this work in progress many times and in many places in the early morning, usually along some roadside or open place, as seen in Fig. 52, but never later in the day. When the warp is laid each will be rolled upon its stretcher and removed to the house to be woven.

In many places in Kiangsu province batteries of the large dye pits were seen sunk in the fields and lined with cement. These were 6 to 8 feet in diameter and 4 to 5 feet deep. In one case there were nine pits in the set. Some of the pits were neatly sheltered beneath live arbours, as represented in Fig. 53. But much of this spinning, weaving, dyeing and printing of late years is being displaced by the cheaper calicos of foreign make, and most of the

Fig. 52. – Laying warp in the country for four bolts of cotton cloth.

dye pits we saw were not now used for this purpose, the two in the illustration serving as manure receptacles. Our interpreter stated, however, that there is a growing dissatisfaction with foreign goods on account of their lack of durability; and we saw many cases where the cloth dyed blue was being dried in large quantities on the grave lands.

In another home for nearly an hour we observed a method of beating cotton and of laying it to serve as the body for mattresses and the coverlets for beds. This we could do without intrusion because the home was also the workshop and opened full width directly upon the narrow street. The heavy wooden shutters which closed the home at night were serving as a work bench about 7 feet square, laid upon movable supports. There was barely room to work between it and the sidewalk without impeding traffic, and on the three other sides there was a floor space 3 or 4 feet wide. In the rear sat grandmother and wife, while in and out the four younger children were playing. Occupying the two sides of the room were receptacles filled with raw cotton and appliances for the work. There may have been a kitchen and sleeping-room behind, but no door, as such, was visible. The finished mattresses, carefully rolled and wrapped in paper, were suspended from the ceiling. On the improvised work-table, with its top 2 feet above the floor, there had been laid in the morning before our visit, a mass of soft white cotton more than 6 feet square and fully 12 inches deep. On opposite sides of this table the father and his son,

FIG. 53. – Two dye pits under woven arbour shelter, now abandoned for their original purpose and used as manure receptacles. The trees in the rear are a typical clump of bamboo so frequently seen about farm-houses.

of twelve years, each twanged the string of their heavy bamboo bows, snapping the lint from the wads of cotton and flinging it broadcast in an even layer over the surface of the growing mattress, the two strings the while emitting tones pitched far below the hum of the bumble-bee. The heavy bow was steadied by a cord secured around the body of the operator, allowing him to manage it with one hand and to move readily round his work. By this means the lint was expeditiously plucked and skilfully and uniformly laid.

Repeatedly, taken in small bits from the barrel of cotton, the lint was distributed over the entire surface with great dexterity and uniformity, the mattress growing upward with perfectly vertical sides, straight edges and square corners. In this manner a thoroughly uniform texture is secured which compresses into a body of even thickness, free from hard places.

The next step in building the mattress is even more simple and expeditious. A basket of long bobbins of roughly spun cotton was near the grandmother and probably her handiwork. The father took from the wall a slender bamboo rod like a fish-pole, 6 feet long, and selecting one of the spools, threaded the strand through an eye in the small end. With the pole and spool in one hand and the free end of the thread, passing through the eye, in the other, the father reached the thread across the mattress to the boy who hooked his finger over it, carrying it to one edge of the bed of cotton. While this was doing the father had whipped the pole back to his side and caught the thread over his own finger, bringing this down upon the cotton opposite his son. There was thus laid a double strand, but the pole continued whipping back and forth across the bed, father and son catching the threads and bringing them to place on the cotton at the rate of forty to fifty courses per minute, and in a very short time the entire surface of the mattress had been laid with double strands. A heavy bamboo roller was next laid across the strands at the middle, passed carefully to one side, back again to the middle and then to the other edge. Another layer of threads was then laid diagonally and this similarly pressed with the same roller; then another diagonally the other way and finally straight across in both directions. A similar network of strands had been laid upon the table before spreading the cotton. Next a flat-bottomed, circular, shallow basket-like form 2 feet in diameter was used to gently compress the material from 12 to 6 inches in thickness. The woven threads were now turned over the edge of the mattress on all sides and sewed down, after which, by means of two heavy solid wooden disks 18 inches in diameter, father and son compressed the cotton until the thickness was reduced to 3 inches. There remained the task of carefully folding and wrapping the finished piece in oiled paper and of suspending it from the ceiling.

On March 20th, when visiting the Boone Road and Nanking Road markets in Shanghai, we had our first surprise regarding the extent to which vegetables enter into the daily diet of the Chinese. We had observed long processions of wheelbarrow men moving from the canals through the streets carrying large loads of the green tips of rape in bundles 1 foot long and 5 inches in diameter. These had come from the country on boats each carrying tons of the succulent leaves and stems. We had counted as many as fifty wheelbarrow men passing a given point on the street in quick succession, each carrying 300 to 500 pounds of the green rape and moving so rapidly that it was not easy to keep pace with them, as we learned in following one of the trains for twenty minutes to its

Fig. 54. – 'Salted cabbage,' prepared from young rape, displayed for sale in Boone Road market, Shanghai.

destination. During this time not a man in the train halted or slackened his pace.

This rape is very extensively grown in the fields, the tips of the stems cut when tender and eaten, after being boiled or steamed, after the manner of cabbage. Very large quantities are also packed with salt in the proportion of about 20 pounds of salt to 100 pounds of the rape. This (Fig. 54) and many other vegetables are sold thus pickled and used as relishes with rice, which invariably is cooked and served without salt or other seasoning.

Another field crop very extensively grown for human food, and partly as a source of soil nitrogen, is closely allied to our alfalfa. This is the *Medicago astragalus*, two beds of which are seen in Fig. 55. Tender tips of the stems are gathered before the stage of blossoming is reached and served as food after boiling or steaming. It is known among the foreigners as Chinese 'clover.' The stems are also cooked and then dried for use when the crop is out of season. When picked *very* young, wealthy Chinese families pay an extra high price for the tender shoots, sometimes as much as 20 to 28 cents, U.S. currency, per pound.

The markets are thronged with people making their purchases in the early mornings, and the congested condition, with the great variety of vegetables, makes it almost as impressive a sight as Billingsgate fish market in London. In the following table we give a list of vegetables observed there and the prices at which they were selling.

LIST OF VEGETABLES DISPLAYED FOR SALE IN BOONE ROAD MARKET, SHANGHAI, APRIL 6TH, 1909, WITH PRICES EXPRESSED IN U.S. CURRENCY.

	Cents.		Cents.
Lotus roots, per lb.	1·60	Maize, shelled, per lb.	1·00
Bamboo sprouts, per lb.	6·40	Windsor beans, dry, per lb.	1·72
English cabbage, per lb.	1·33	French lettuce, per head	·44
Olive greens, per lb.	·67	Hau Tsai, per head	·87
White greens, per lb.	·33	Cabbage lettuce, per head	·22
Tee Tsai, per lb.	·53	Kale, per lb.	1·60
Chinese celery, per lb.	·67	Rape, per lb.	·23
Chinese clover, per lb.	·53	Portuguese watercress, per basket	2·15
Chinese clover, very young, per lb.	21·33	Shang tsor, per basket	8·60
Oblong white cabbage, per lb.	2·00	Carrots, per lb.	·97
Red beans, per lb.	1·33	String beans, per lb.	1·60
Yellow beans, per lb.	1·87	Irish potatoes, per lb.	1·60
Peanuts, per lb.	2·49	Red onions, per lb.	4·96
Ground nuts, per lb.	2·96	Long white turnips, per lb.	·44
Cucumbers, per lb.	2·58	Flat string beans, per lb.	4·80
Green pumpkin, per lb.	1·62	Small white turnips, bunch	·44

	Cents.		Cents.
Onion stems, per lb.	1·29	Large sweet potatoes, per lb.	1·33
Lima beans, green, shelled per lb.	6·45	Small sweet potatoes, per lb.	1·00
Egg plants, per lb.	4·30	Onion sprouts, per lb.	2·13
Tomatoes, per lb.	5·16	Spinach, per lb.	1·00
Small flat turnips, per lb.	·86	Fleshy stemmed lettuce, peeled, per lb.	2·00
Small red beets, per lb.	1·29	Fleshy stemmed lettuce, unpeeled, per lb.	·67
Artichokes, per lb.	1·29	Bean curd, per lb.	3·93
White beans, dry, per lb.	4·30	Shantung walnuts, per lb.	4·30
Radishes, per lb.	1·29	Duck eggs, per dozen	8·34
Garlic, per lb.	2·15	Hen's eggs, per dozen	7·30
Kohl rabi, per lb.	2·15	Goat's meat, per lb.	6·45
Mint, per lb.	4·30	Pork, per lb.	6·88
Leeks, per lb.	2·13	Hens, live weight, per lb.	6·45
Large celery, bleached, per bunch	2·10	Ducks, live weight, per lb.	5·59
Sprouted peas, per lb.	·80	Cockerels, live weight, per lb.	5·59
Sprouted beans, per lb.	·93		
Parsnips, per lb.	1·29		
Ginger roots, per lb.	1·60		
Water chestnuts, per lb.	1·33		

This long list, made up chiefly of fresh vegetables displayed for sale on one market day, is by no means complete. The record is only such as was made in passing down one side and across one end of the market occupying nearly one city block. Nearly everything is sold by weight and the problem of correct weights is effectively solved by each purchaser carrying his own scales, which he unhesitatingly uses in the presence of the dealer. These scales are made on the pattern of the old time steelyards, but from slender rods of wood or bamboo provided with a scale and sliding poise, the suspensions all being made with strings.

We stood by through the purchasing of two cockerels and the dickering over their weight. A dozen live birds were under cover in a large openwork basket. The customer took out the birds one by one, examining them by touch, finally selecting two, the price being named. These the dealer tied together by their feet and weighed them, announcing the result; whereupon the customer checked the statement with his own scales. An animated dialogue followed, punctuated with many gesticulations and with the customer tossing the birds into the basket and turning to go away while the dealer grew more earnest. The purchaser finally turned back, and again balancing the roosters upon his scales, called a bystander to read the weight, and then flung them in apparent disdain at the dealer, who caught them and placed them in the customer's basket. The storm subsided and the dealer accepted

Fig. 55. – Two beds of Chinese clover (*Medicago astragalus*) grown in the garden for human food in the season and for soil fertility later.

92c. Mexican, for the two birds. They were good-sized roosters and must have dressed more than three pounds each, yet for the two he paid less than 40 cents, U.S. currency.

Bamboo sprouts are very generally used in China, Korea and Japan, and when one sees them growing they suggest giant stalks of asparagus, some of them being 3 and even 5 inches in diameter and a foot in height at the stage for cutting. They are shipped in large quantities to provinces where they do not grow or when they are out of season. Those we saw in Nagasaki had come from Canton or Swatow or possibly Formosa. The form and foliage of the bamboo give the most beautiful effects in the landscape, especially when grouped with tree forms. They are usually cultivated in small clumps about dwellings in places not otherwise readily utilized, as seen in Fig. 53. Like the asparagus bud, the bamboo sprout grows to its full height between April and August, even when it exceeds 30 or even 60 feet in height. The buds spring from fleshy underground stems or roots whose stored nourishment permits rapid growth, which in its earlier stages will exceed 12 inches in twenty-four hours. But while the full size of the plant is attained the first season, three or four years are required to ripen and harden the wood sufficiently to make it suitable for the many uses to which the stems are put.

Lotus roots form another article of diet largely used and widely cultivated from Canton to Tokio. These are seen in the lower section of Fig. 56, and the plants in bloom in Fig. 57, growing in water, their natural habitat. The lotus is grown in permanent ponds not readily drained for rice or other crops, and the roots are widely shipped.

Sprouted beans and peas of many kinds and the sprouts of other vegetables, such as onions, are very generally seen in the markets of both China and Japan, at least during the late winter and early spring, and are sold as foods.

Ginger is another crop which is extensively cultivated. It is generally displayed in the market in the root form. No one thing was more generally hawked about the streets of China than the water chestnut. This is a small corm or fleshy bulb having the shape and size of a small onion. Boys pare them and sell a dozen spitted together on slender sticks the length of a knitting-needle. Then there are the water caltropes, grown in the canals, producing a fruit resembling a horny nut having a shape which suggests for them the name 'buffalo-horn.' Still another plant, known as water-grass (*Hydropyrum latifolium*), is grown in Kiangsu province, where the land is too wet for rice. The plant has a tender succulent crown of leaves and the peeling of the outer coarser ones away suggests the husking of an ear of green corn. The portion eaten is the central tender new growth, and when cooked forms a delicate savoury dish. The farmers' selling price is 3 to 4 dollars, Mexican, per 100 catty, or $·97 to $1·29 per hundredweight, and the return per acre is from $13 to $20.

The small number of animal products which are included in the market list given should not be taken as indicating the proportion

FIG. 56. – Boone Road vegetable market, April 6th, Shanghai, China. The large vegetables in the lower section are lotus roots.

of animal to vegetable foods in the dietaries of these people. It is nevertheless true that they are vegetarians to a far higher degree than are most Western nations, and the high maintenance efficiency of the agriculture of China, Korea and Japan is in great measure rendered possible by the adoption of a diet so largely vegetarian. Hopkins, in his *Soil Fertility and Permanent Agriculture*, page 234, makes this pointed statement of fact: '1,000 bushels of grain has at least five times as much food value and will support five times as many people as will the meat or milk that can be made from it.' He also calls attention to the results of many

Fig. 57. – Lotus pond with plant in bloom; cultivated for their fleshy roots used for food, shown in Fig. 56.

Rothamsted feeding experiments with growing and fattening cattle, sheep and swine, showing that the cattle destroyed outright 57·3 pounds in every 100 pounds of dry substance eaten, and that this passes off into the air, as the whole of wood does except the ashes, when burned in the stove. They left in the excrements 36·5 pounds, and stored as increase only 6·2 pounds out of the 100. With sheep the corresponding figures were 60·1 pounds; 31·9 pounds and 8 pounds; and with swine they were 65·7 pounds; 16·7 pounds and 17·6 pounds. But less than two-thirds of the substance stored in the animal can become food for man. Hence we get only 4 pounds in 100 of the dry substances eaten by cattle in the form of human food; only 5 pounds from the sheep; and 11 pounds from swine.

In view of these relations, recently established as scientific facts by rigid research, it is remarkable that these ancient people came long ago to discard cattle as milk and meat producers; to use sheep more for their pelts and wool than for food; while swine are the one kind of the three classes which they retained in the rôle of middleman as transformers of coarse substances into human food.

It is clear that in the adoption of the succulent forms of vegetables as human food important advantages are gained. At this stage of maturity they have a higher digestibility, thus making the elimination of the animal less difficult. Their nitrogen content is relatively higher and this in a measure compensates for loss of meat. By devoting the soil to growing vegetation which man can directly digest they have saved 60 pounds per 100 of absolute waste by the animal, returning their own wastes to the field for the maintenance of fertility. In using these immature forms of vegetation so largely as food they are able to produce an immense amount that would otherwise be impossible, as it is grown in a shorter time and the same soil produces more crops. It is also produced late in the fall and early in the spring when the season is too cold and the hours of sunshine too few each day to permit of ripening crops.

VII

THE FUEL PROBLEM, BUILDING AND TEXTILE MATERIALS

With the vast and ever increasing demands made upon the products of cultivated fields, for food, for apparel, for furnishings and for cordage, better soil management must grow more important as populations multiply. With the increasing cost and ultimate exhaustion of mineral fuel; with our timber vanishing rapidly before the ever-growing demands for lumber and paper; with the inevitably slow growth of trees and the very limited areas which the world can ever afford to devote to forestry, the time must surely come when, in short period rotations, there will be grown upon the farm materials from which to manufacture not only paper and the substitutes for lumber, but fuels as well. The complete utilization of every stream which reaches the sea, reinforced by the force of the winds and the energy of the waves, cannot fully meet the demands of the future for power and heat; hence only in the event of science and engineering skill becoming able to transform the unlimited energy of space through which we are whirled, with an economy approximating that which crops now exhibit, can good soil management be relieved of the task of meeting a portion of the world's demand for power and heat.

When these statements were made in 1905 we did not know that for centuries there had existed in China, Korea and Japan a density of population such as to require the extensive cultivation of crops for fuel and building material, as well as for fabrics, by the ordinary methods of tillage, and hence another of the many surprises we had was the solution these people had reached of their fuel problem. Their solution is direct and the simplest possible. Dress to make fuel for warmth of body unnecessary, and burn the coarser stems of crops, such as cannot be eaten, or employed to feed animals or otherwise made useful. These people still use what wood can be grown on the untillable land within transporting distance, and convert much wood into charcoal, making transportation over longer distances easier. The general use of mineral fuels, such as coal, coke, oils and gas, had been impossible to these as to every other people until within the last one hundred years. Coal, coke, oil and natural gas, however, have been locally used by the Chinese from very ancient times. For more than two thousand years brine from many deep wells in Szechwan province has been evaporated with heat generated by the burning of natural gas from wells, conveyed through bamboo stems to the pans and burned from iron terminals. In other sections of the same province much brine is evaporated over coal fires. Alexander Hosie estimates the production of salt in Szechwan province at more than 600 million pounds annually.

Coal is here used also to some extent for warming the houses, burned in pits sunk in the floor, the smoke escaping where it may. The same method of heating we saw in use in the post office at Yokohama during February. The fires were in large iron braziers

Fig. 58. – Charcoal balls briquetted with rice water or clay, for use as fuel.

more than 2 feet across the top, simply set about the room, three being in operation.

In both China and Japan we saw coal dust put into the form and size of medium oranges by mixing it with a thin paste of clay. Charcoal is similarly moulded, as seen in Fig. 58, a by-product from the manufacture of rice syrup being used for cementing. In Nanking we watched with much interest the manufacture of charcoal briquettes by another method. A Chinese workman was seated upon the earth floor of a shop. By his side was a pile of powdered charcoal, a dish of rice syrup by-product and a basin of the moistened charcoal powder. Between his legs was a heavy mass of iron containing a slightly conical mould 2 inches deep, $2\frac{1}{2}$ inches across at the top, and a heavy iron hammer weighing several pounds. In his left hand he held a short heavy ramming tool and with his right placed in the mould a pinch of the moistened charcoal: then followed three well-directed blows from the hammer upon the ramming tool, compressing the charge of moistened, sticky charcoal into a very compact layer. Another pinch of charcoal was added and the process repeated until the mould was filled, when the briquette was forced out.

By this simplest possible mechanism, the man, utilizing but a small part of his available energy, was subjecting the charcoal to an enormous pressure, such as we attain only with the best hydraulic presses. He was using the principle of repeated small charges recently patented and applied in our large and most efficient cotton and hay presses—a process which permits much denser bales to be made than is possible when large charges are added. The Chinese is here, as in a thousand other ways, thoroughly sound in his application of mechanical principles.

A Shantung farmer in winter dress (Fig. 16) and the Kiangsu woman portrayed in Fig. 59, in corresponding costume, are typical illustrations of the manner in which food for body warmth is minimized and of the way the heat generated in the body is conserved. Observe the farmer's wadded and quilted frock, his trousers of similar goods tied about the ankle, with his feet clad in multiple socks and cloth shoes provided with thick felted soles. These types of dress, with the wadding, quilting, belting and tying, incorporate and confine as part of the effective material a large volume of air, thus securing without cost, much additional warmth without increasing the weight of the garments. Beneath these outer garments several under-pieces of different weights are worn, which greatly conserve the warmth during the coldest weather and make possible a wide range of adjustment to suit varying changes in temperature. It is doubtful if there could be devised a wardrobe suited to the conditions of these people at a smaller first cost and maintenance expense. Rev. E. A. Evans, of the China Inland Mission, for many years residing at Sunking in Szechwan, estimated that a farmer's wardrobe, once it was procured, could be maintained with an annual expenditure of $2·25 of our currency, this sum procuring the materials for both repairs and renewals.

The intense individual economy, extending to the smallest matters, so universally practised by these people, has sustained the massive strength of the Mongolian nations through their long history, and this trait is seen in their handling of the fuel problem, as it is in all other lines. In the home of Mrs. Wu, owner and manager of a 25-acre rice farm in Chekiang province, there was a masonry kang 7 by 7 feet, about 28 inches high, which could be warmed in winter by building a fire within. The top was fitted for mats to serve as couch by day and as a place upon which to

FIG. 59. – A Kiangsu country woman in winter dress.

spread the bed at night. In the Shantung province we visited the home of a prosperous farmer and here found two kangs in separate sleeping-apartments, both warmed by the waste heat from the kitchen, whose chimney flue passed horizontally under the kangs before rising through the roof. These kangs were wide enough to spread the beds upon, about 30 inches high. They were constructed of brick 12 inches square and 4 inches thick. The bricks were made from the clay subsoil taken from the fields, worked into a plastic mass, mixed with chaff and short straw, dried in the sun, and then laid in a mortar of the same material. These massive kangs are thus capable of absorbing large amounts of the waste heat from the kitchen during the day, and this is again given out in congenial warmth by day and to the beds and sleeping-apartments during the night. In some Manchurian inns large compound kangs are so arranged that the guests sleep heads together in double rows, separated only by low dividing rails. The greatest economy of fuel is thus secured, and the guests are provided with places where they may sit upon the moderately warmed fire-place, and spread their beds when they retire.

The economy of the chimney beds does not end with the warmth conserved. The earth and straw bricks through the processes of fermentation and through shrinkage, become open and porous after three or four years of service, causing the draught to be defective and giving annoyance from smoke, so that renewal becomes necessary. But the heat, the fermentation and the absorption of products of combustion have together transformed the comparatively infertile subsoil into what they regard as a valuable fertilizer, and these discarded bricks are used in the preparation of compost fertilizers for the fields.

Our own observations have shown that heating soils to dryness at a temperature of 110° C. greatly increases the freedom with which plant food may be recovered from them by the solvent power of water, and the same heating doubtless improves the physical and biological conditions of the soil as well. Nitrogen combined as ammonia, and phosphorus, potash and lime are all carried with the smoke or soot filter into the porous brick, and thus add plant food directly to the soil. Soot from wood has been found to contain, on an average, 1·36 per cent of nitrogen; ·51 per cent of phosphorus, and 5·34 per cent of potassium. We practise burning straw and corn-stalks in enormous quantities, to get them easily out of the way, thus scattering on the winds valuable plant food, thoughtlessly and lazily wasting where these people laboriously and religiously save. These are gains in addition to those which result from the formation of nitrates, soluble potash and other plant foods through fermentation. We saw many instances where these discarded bricks were being used, both in Shantung and Chihli provinces, and it was common in walking through the streets of country villages to see piles of them, evidently recently removed.

The fuel grown on the farms consists of the stems of all agricultural crops which are to any extent woody, unless they can be put to some better use. Rice straw, cotton stems pulled by the roots after the seed has been gathered, the stems of windsor beans, those of rape and the millets, all pulled by the roots, and many other kinds, are brought to the market tied in bundles in the manner seen in Figs. 60, 61 and 62. These fuels are used for domestic purposes and for the burning of lime, brick, roofing tile and earthenware, as well as in the manufacture of oil, tea, bean-curd and many other processes. In the home, when the meals are cooked with these light bulky fuels, it is the duty of someone, often one of the children, to sit on the floor and feed the fire with one hand while with the other a bellows is worked to secure sufficient draught.

The manufacture of cotton-seed oil and cotton-seed cake is one of the common family industries in China, and in one of these homes we saw rice hulls and rice straw being used as fuel. In the large, low, one-story, tile-roofed building serving as shop, warehouse, factory and dwelling, a family of four generations were at work, the grandfather supervising in the mill and the grandmother leading in the home and shop where the cotton-seed oil was being retailed for 22 cents, gold, per pound and the cotton-seed cake at 33 cents per hundredweight. Back of the shop and living-rooms, in the mill compartment, three blindfolded water buffalo, each working a granite mill, were crushing and grinding the cotton-seed. Three other buffalo, for relay service, were lying at rest or eating, awaiting their turn at the ten-hour working day. Two of the mills were horizontal granite burrs more than 4 feet in diameter, the upper one revolving once with each circuit made by the cow. The third mill was a pair of massive granite rollers, each 5 feet in diameter and 2 feet thick, joined on a very short horizontal axle which revolved on a circular stone plate about a vertical axis once with each circuit of the buffalo. Two men tended the three mills. After the cotton-seed had been twice passed through the mills it was steamed to render the oil fluid and more readily

FIG. 60. – Boat-loads of fuel, mainly bundles of rice straw and cotton stems, on Soochow creek, Shanghai.

expressed. The steamer consisted of two covered wooden hoops provided with screen bottoms, and in these the meal was placed over openings in the top of an iron kettle of boiling water from which the steam was forced through the charge of meal. Each charge was weighed in a scoop balanced on the arm of a bamboo scale, and so a uniform weight for the cakes was secured.

On the ground in front of the furnace sat a boy of twelve years steadily feeding a fire with rice chaff with his left hand at the rate

Fig. 61. – Cotton-stem fuel being conveyed from the canals to city market stalls.

of about thirty charges per minute, while with his right hand, and in perfect rhythm, he drew back and forth the long plunger of a rectangular box bellows, maintaining a forced draught for the fire. At intervals the man who was bringing fuel fed the furnace with a bundle of rice straw, thus giving the boy's left arm a moment's respite. When the steaming had rendered the oil sufficiently fluid the meal was transferred, hot, to 10-inch hoops 2 inches deep, made of braided bamboo strands, and deftly tramped with the bare feet, while hot, the operator steadying himself with a pair of hand

FIG. 62. – Rice straw fuel being conveyed from canal boats to city market stalls.

bars. After a stack of sixteen hoops, divided by a slight sifting of chaff or short straw to separate the cakes, had been completed, these were taken to one of four pressmen, who were kept busy in expressing the oil.

The presses consisted of two parallel timbers framed together, long enough to receive the sixteen hoops on edge above a gap between them. These cheeses of meal were subjected to an enormous pressure secured by means of three parallel lines of wedges forced against the follower each by an iron-bound master wedge, driven home with a heavy beetle weighing some 25 or 30 pounds. The lines of wedges were tightened in succession, the loosened line receiving an additional wedge to take up the slack after drawing back the master wedge, which was then driven home. To keep good the supply of wedges, which are often crushed under the pressure, a second boy, older than the one at the furnace, was working on the floor, shaping new ones, the broken wedges and the chips going to the furnace for fuel.

By this very simple, readily constructed and inexpensive mechanism enormous pressures were secured and when the operator had obtained the desired compression he lighted his pipe and sat down to smoke until the oil ceased dripping into the pit sunk in the floor beneath the press. In this interval the next series of cakes went to another press and the work was thus kept up throughout the day. Six hundred and forty cakes was the average daily output of this family of eight men and two boys, with their six water buffalo.

The cotton-seed cakes were being sold as feed, and a near-by Chinese dairyman was using them for his herd of forty water buffalo, seen in Fig. 63, producing milk for the foreign trade in Shanghai. This herd of forty cows, one of which was an albino, was giving an average of but 200 catty of milk per day, or at the rate of $6\frac{2}{3}$ pounds per head! The cows have extremely small udders but the milk is very rich, as indicated by an analysis made in the office of the Shanghai Board of Health and obtained through the kindness of Dr. Arthur Stanley. The milk showed a specific gravity of 1·028 and contained 20·1 per cent total solids; 7·5 per cent fat; 4·2 per cent milk sugar and ·8 per cent ash. In the family of the Rev. W. H. Hudson, of the Southern Presbyterian Mission, Kashing, whose very gracious hospitality we enjoyed on two different occasions, the butter made from the milk of two of these cows, one of which, with her calf, is seen in Fig. 64, was used on the family table. It was as white as lard or cottolene, but the texture and flavour were normal and far better than the Danish and New Zealand products served at the hotels.

The milk produced at the Chinese dairy in Shanghai was being sold in bottles holding 2 pounds, at the rate of $1 a bottle, or 43

Fig. 63.—A dairy herd of water buffalo owned by a Chinese farmer who was supplying milk to foreigners in Shanghai.

cents, gold. This seems high and there may have been misunderstanding on the part of my interpreter, but his answer to my question was that the milk was being sold at one Shanghai dollar per bottle holding one and a half catty, which, interpreted, is the value given above.

But fuel from the stems of cultivated plants which are in part otherwise useful, is not sufficient to meet the needs of country and village, notwithstanding the intense economies practised. Large areas of hill and mountain land are made to contribute their share, as we have seen in the south of China, where pine boughs were being used for firing the lime and cement kilns. At Tsingtao we saw the pine bough fuel on the backs of mules (Fig. 65), coming from the hills in Shantung province. Similar fuels were being used in Korea and we have photographs of large pine bough fuel stacks, taken in Japan at Funabashi, east from Tokyo.

The hill and mountain lands, wherever accessible to the densely peopled plains, have long been cut over and as regularly has afforestation been encouraged and deliberately secured even through the transplanting of nursery stock grown expressly for

FIG. 64. – Water buffalo and calf, Kashing, Chekiang province.

that purpose. We had read so much regarding the reckless destruction of forests in China and Japan and had seen so few old forest trees except where these had been protected about temples, graves or houses, that when Rev. R. A. Haden, of the Elizabeth Blake hospital, near Soochow, insisted that the Chinese were deliberate foresters and that they regularly grow trees for fuel, transplanting them when necessary to secure a close and early stand, after the area had been cleared, we were so much surprised that he generously volunteered to accompany us westward on a two days' journey into the hill country where the practice could be seen.

A family owning a houseboat and living upon it was engaged for the journey. This family consisted of a recently widowed father, his two sons, newly married, and a helper. They were to transport us and provide sleeping quarters for myself, Mr. Haden and a cook for the consideration of $3·00, Mexican, per day and to continue the journey through the night, leaving the day for observation in the hills.

The recent funeral had cost the father $100 Mex., the wedding of the two sons $50 each, while the remodelling of the house-boat to meet the needs of the new family relations cost still another

FIG. 65. – Pine bough fuel coming into Tsingtao from the Shantung hills.

$100. To meet these expenses it had been necessary to borrow the full amount, $300. On $100 the father was paying 20 per cent interest; on $50 he was compelled to pay 50 per cent interest. The balance he had borrowed from friends without interest but with the understanding that he would return the favour should occasion be required.

Rev. A. E. Evans informed us that it is a common practice in China for neighbours to help one another in times of great financial stress. This is one of the methods: A neighbour may need 8,000 cash. He prepares a feast and sends invitations to a hundred friends. They know there has been no death in his family and that there is no wedding, still it is understood that he is in need of money. The feast is prepared at a small expense, the invited guests come, each bringing eighty cash as a present. The recipient is expected to keep a careful record of contributing friends and to repay the sum. Another method is like this: For some reason a

FIG. 66. – Residence houseboat used by family for carrying passengers on rivers and canals, China.

man needs to borrow 20,000 cash. He proposes to twenty of his friends that they organize a club to raise this sum. If the friends agree each pays 1,000 cash to the organizing member. The balance of the club draw lots as to which member shall be number two, three, four, five, etc., designating the order in which repayments shall be made. The man borrowing the money is then under obligation to see that these payments are paid in full at the times agreed upon. Not infrequently a small rate of interest is charged.

Rates of interest are very high in China, especially on small sums where securities are not the best. Mr. Evans informs me that 2 per cent per month is low, and 30 per cent per annum is very commonly collected. Such obligations are often never met, but they do not outlaw and may descend from father to son.

The boat cost $292.40 in U.S. currency; the yearly earning was $107.50 to $120.40. The funeral cost $43, and $43 more was required for the wedding of the two sons. They were receiving for the services of six people $1·29 per day. An engagement for two weeks or a month could have been made for materially lower rates and their average daily earning, on the basis of 300 days service in the year, and the $120.40 total earning, would be only 40.13 cents, less than 7 cents each. Hence their trip with us was two of their banner days. Foreigners in Shanghai and other cities frequently engage such houseboat service for two weeks or a month of travel on the canals and rivers, finding it a very enjoyable as well as inexpensive way of having a picnic outing.

On reaching the hill lands the next morning there were such scenes as shown in Fig. 67, where the strips of tree growth, varying from two to ten years, stretched directly up the slope, often in strong contrast on account of the straight boundaries and different ages of the timber. Some of these long narrow holdings were less than 2 rods wide and on one of these only recently cut, up which we walked for a considerable distance, the young pine were springing up in goodly numbers. As many as eighteen young trees were counted on a width of 6 feet across the strip of 30 feet wide. On this area everything had been recently cut clean. Even stumps and the large roots were dug and saved for fuel.

In Fig. 68 are seen bundles of fuel from such a strip, just brought into the village, the boughs retaining the leaves although the fuel had been dried. The roots, too, are tied in with the limbs so that everything is saved. On our walk to the hills we passed many people bringing their loads of fuel swinging from carrying poles on their shoulders.

Inquiries regarding the afforestation of these strips of hillside showed that the extensive digging necessitated by the recovery of the roots usually caused new trees to spring up quickly as

Fig. 67. – Forest cutting in narrow strips on steep hillsides west of Soochow.

Fig. 68. – Bundles of pine and oak bough fuel gathered on the hill lands west of Soochow, Kiangsu.

volunteers from scattered seed and from the roots, so that planting was not generally required. Talking with a group of people as to where we could see some of the trees used for replanting the hill-sides, a lad of seven years was first to understand and volunteered to conduct us to a planting. This he did and was overjoyed on receipt of a trifle for his services. One of these little pine nurseries is seen in Fig. 69, many being planted in suitable places through the woods. The lad led us to two such locations with whose

FIG. 69. – Tiny nursery of small pines growing among ferns in a shady wood, for replanting cut-over hillsides.

whereabouts he was evidently very familiar, although they were considerable distances from the path and far from home. These small trees are used in filling in places where the volunteer growth has not been sufficiently close. A strong herbaceous growth usually springs up quickly on these newly cleared lands and this too is cut for fuel or for use in making compost or as green manure.

The grass which grows on the grave lands, if not fed off, is also cut and saved for fuel. We saw several instances of this outside Shanghai, one where a mother with her daughter, provided with rake, sickle, basket and bag, were gathering the dry stubble and grass of the previous season from the grave lands where there was less than could be found on our closely mowed meadows. In Fig. 70 may be seen a man who has just returned with such a load, and in his hand is the typical rake of the Far East, made by simply bending bamboo splints, claw-shape, and securing them as seen in the engraving.

FIG. 70.—Dried grass fuel gathered on grave lands, Shanghai.

In the Shantung province, in Chihli and in Manchuria, millet stems, especially those of the great kaoliang or sorghum, are extensively used for fuel and for building as well as for screens, fences and matting. At Mukden the kaoliang was selling as fuel at $2.70 to $3.00, Mexican, for a 100-bundle load of stalks, weighing 7 catty to the bundle. The yield per acre of kaoliang fuel amounts to 5,600 pounds and the stalks are 8 to 12 feet long, so that when carried on the backs of mules or horses the animals are nearly hidden by the load. The price paid for plant stem fuel from agricultural crops, in different parts of China and Japan, ranged from $1.30 to $2.85, U.S. currency, per ton. The price of anthracite coal at Nanking was $7.76 per ton. Taking the weight of dry oak wood at 3,500 pounds per cord, the plant stem fuel, for equal weight, was selling at $2.28 to $5.00.

FIG. 71. – Bundles of kaoliang fuel coming into Kiaochow market, Shuntung.

Large amounts of wood are converted into charcoal in these countries and sent to market baled in rough matting, or in basket-work cases woven from small brush and holding 2 to 2½ bushels. When such wood is not converted into charcoal it is sawed into 1- or 2-foot lengths, split and marketed tied in bundles.

Along the Mukden-Antung railway in Manchuria fuel was also being shipped in 4-foot lengths, in the form of cordwood. In Korea cattle were provided with a peculiar saddle for carrying wood in 4-foot sticks laid blanket-fashion over the animal, extending far down on their sides. As in most parts of China where we visited, the tree growth over the hills was generally scattering and thin on the ground wherever there was not individual ownership in small holdings. Under and among the scattering pine there were oak in many cases, but these were always small, evidently of not more than two or three years' standing, and appearing to have been repeatedly cut back. It was in Korea that we saw so many instances of young leafy oak boughs brought to the rice fields and used as green manure.

There was abundant evidence of periodic cutting between Mukden and Antung in Manchuria; between Wiju and Fusan in Korea; and throughout most of our journey in Japan – from Nagasaki to Moji and from Shimonoseki to Yokohama. In all of these countries afforestation takes place quickly and the cuttings on private holdings are made once in ten, twenty or twenty-five years. When the wood is sold to those coming for it the receivers pay at the rate of 40 sen per one-horse load of 40 kan, or 330 pounds, such as is seen in Fig. 72. Director Ono, of the Akashi Experiment station, informed us that such fuel loads in that prefecture, where the wood is cut once in ten years, bring returns amounting to about $10 per acre, for the ten-year crop. This land was worth $40 per acre, but when they are suitable for orange groves they sell for $600 per acre. Mushroom culture is extensively practised under the shade of some of these wooded areas, yielding under favourable conditions at the rate of $100 per acre.

The forest-covered area in Japan, exclusive of Formosa and Karafuto, amounts to a total of 54,196,728 acres, less than 20,000,000 of which are in private holdings, the balance belonging to the state and to the Imperial Crown.

In all these countries there has been an extensive general use of materials other than wood for building purposes; and very many of the substitutes for lumber are products grown on the cultivated fields. The use of rice straw for roofing, as seen in the Hakone village, Fig. 7, is very general throughout the rice-growing districts, and even the sides of houses may be similarly thatched, as was observed in the Canton delta region, such a construction being warm for winter and cool for summer. The life of these thatched roofs, however, is short and they must be renewed as often as every three to five years, but the old straw is highly prized as fertilizer for the fields on which it is grown, or it may serve as fuel, the ashes only going to the fields.

Burned clay tile, especially for the cities and public buildings, are very extensively used for roofing, clay being abundant and near at hand. In Chihli and in Manchuria millet and sorghum

Fig. 72. – Japanese fuel coming down from the wooded hills.

stems, used alone or plastered, as in Fig. 73, with a mud mortar, sometimes mixed with lime, cover the roofs of vast numbers of the dwellings outside the larger cities.

At Chiao Tou in Manchuria we saw the building of the thatched millet roofs and the use of kaoliang stems instead of timber. Rafters were set in the usual way and covered with a layer about 2 inches thick of the long kaoliang stems stripped of their leaves and tops. These were tied together and to the rafters with twine, thus forming a sort of matting. A layer of thin clay mortar was then spread over the surface and well trowelled until it began to show on the under side. Over this was applied a thatch of small millet stems bound in bundles 8 inches thick, cut square across the butts to 18 inches in length. They were dipped in water and laid in courses after the manner of shingles, but the butts of the stems were driven forward to a slope which obliterated the shoulder, making the courses invisible. In the better houses this thatching may be plastered with earth mortar or with an earth-lime mortar, which is less liable to wash in heavy rain.

The walls of the house we saw building were also sided with the long, large kaoliang stems. An ordinary frame with posts

Fig. 73. – Millet-thatched roofs plastered with earth; mud chimneys; walls of houses plastered with mud, and winter storage pits for vegetables built of clay and chaff mortar.

and girts about 3 feet apart had been erected, on sills and with plates carrying the roof. Standing vertically against the girts and tied to them, forming a close layer, were the kaoliang stems. These were plastered outside and in with a layer of thin earth mortar. A similar layer of stems, set up on the inside of the girts and similarly plastered, formed the inner face of the wall of the house, leaving dead air spaces between the girts.

Bricks made from earth and dried in the sun (Fig. 74) are very extensively used for house building, with chaff and short straw as a binding material. A house in the process of building, where bricks of this kind were being used, is seen in Fig. 75. The foundation of the dwelling, it will be observed, was laid with well-formed hard-burnt bricks, these being necessary to prevent capillary moisture from the ground being drawn up and softening the earth brick.

Several kilns for burning brick, built of clay and earth, were passed in our journey up the Pei ho, and stacked about them, covering an area of more than 800 feet back from the river, were bundles of the kaoliang stems to serve as fuel in the kilns.

FIG. 74. – Air-dried earth brick for house building.

The extensive use of the unburnt brick is necessitated by the difficulty of obtaining fuel, and various methods are adopted to reduce the number of burnt bricks required in construction. One of these devices is shown in Fig. 64, where the city wall surrounding Kashing is constructed of alternate courses of four layers of burnt bricks separated by layers of simple sun-dried bricks.

In addition to the multiple-function farm-grown crops used for food, fuel and building material, there is a large acreage devoted to the growing of textile and fibre products, of which enormous quantities are produced annually. In Japan, where some 50 millions of people are chiefly fed on the produce of little more than 21,000 square miles of cultivated land, there was grown in 1906 more than 75,500,000 pounds of cotton, hemp, flax and China grass textile stock, occupying 76,700 acres of the cultivated land. On 141,000 acres there grew 115,000,000 pounds of paper mulberry and Mitsumata, materials used in the manufacture of paper. From still another 14,000 acres were taken 92,000,000 pounds of matting stuff, while more than 957,000 acres were occupied by mulberry

Fig. 75. – Foundation of dwelling, consisting of hard-burnt brick; balance of wall to be sun-dried earth brick, seen in Fig. 74.

trees for the feeding of silkworms, yielding to Japan 22,389,798 pounds of silk. Here are more than 300,000,000 pounds of fibre and textile stuff taken from 1,860 square miles of the cultivated land, cutting down the food-producing area to 19,263 square miles; and this area is made still smaller by devoting 123,000 acres to tea, producing in 1906 58,900,000 pounds, worth nearly $5,000,000. Nor do these statements express the full measure of the producing power of the 21,321 square miles of cultivated land, for, in addition to the food and other materials named, $2,365,000 worth of braid were made from straw and wood shavings; $6,000,000 worth of rice straw bags, packing cases and matting; and $1,085,000 worth of wares from bamboo, willow and vine. As illustrating the intense home industry of these people, we may consider the fact that the 5,453,309 households of farmers in Japan produced in 1906, in their homes as subsidiary work, $20,527,000 worth of manufactured articles. If correspondingly exact statistical data were available from China and Korea a similarly full utilization of cultural possibilities would be revealed there.

FIG. 76. – Earth and clay brick kiln on the bank of the Pei-ho, using sorghum stems for fuel.

This marvellous heritage of economy, industry and thrift, bred of the stress of centuries, must not be permitted to lose virility through contact with western wasteful practices, now exalted to seeming virtues through the dazzling brilliancy of mechanical achievements. More and more must labour be dignified in all homes alike, and economy, industry and thrift become inherited impulses, compelling and satisfying.

Cheap, rapid, long-distance transportation, already well started in these countries, will bring with it a fuller utilization of the large stores of coal and mineral wealth and of the enormous available water-power, and as a result there will come some temporary lessening of the stress for fuel and, with better forest management, some relief along the lines of building materials. But the time is not a century distant when, throughout the world, a fuller, better development must take place along the lines of these far-reaching and fundamental practices so long and so effectively followed by the Mongolian races in China, Korea and Japan. When the enormous water-power of these countries has been harnessed and brought into the foot-hills and down upon the margins of the valleys and plains in the form of electric current, let it, if possible, be so distributed as to become available in country village homes to lighten the burden and lessen human drudgery. If this is done the efficiency of the human effort now so well bestowed upon subsidiary manufactures under the guidance and initiative of the home will be increased, and it will be possible for children to grow up to manhood and womanhood under the best conditions possible, rather than in enormous congested factories.

VIII

TRAMPS AFIELD

On March 31st we took the 8 a.m. train on the Shanghai-Nanking railway for Kunshan, situated 32 miles west from Shanghai, to spend the day walking in the fields. The fare, second class, was 80 cents, Mexican. A third-class ticket would have been 40 cents, and a first class $1.60, practically two cents, one cent and half a cent, U.S. currency, per mile. The second-class fare to Nanking, a distance of 193 miles, was $1.72, U.S. currency, or a little less than 1 cent per mile. While the car seats were not upholstered, the service was good. Meals were served on the train in either foreign or Chinese style, and tea, coffee or hot water to drink. Hot, wet face-cloths were regularly passed round and many Chinese daily newspapers were sold on the train.

In the vicinity of Kunshan a large area of farm land had been acquired by the French Catholic Mission at a purchase price of $40, Mexican, per mow, or at the rate of $103.20 per acre.

It was here that we first saw, at close range, the details of using canal mud as a fertilizer, so extensively applied in China. Walking through the fields we came upon the scene illustrated in the middle section of Fig. 77, where, close on the right, was such a reservoir as seen in Fig. 48. Men were in it, dipping up the mud which had accumulated over its bottom, and pouring it on the bank in a field of windsor beans. The thin mud was then over 2 feet deep at that side and flowing into the beans, where it had already spread 2 rods, burying the plants as the engraving shows. When sufficiently dry to be readily handled this would be spread among the beans, as shown in the upper section of the illustration. Here four men were distributing such mud, which had dried, between the rows, not to fertilize the beans, but for a succeeding crop of cotton soon to be planted between the rows, before the latter were harvested. The owner of this piece of land, with whom we talked and who was superintending the work, stated that his usual yield of these beans was 300 catty per mow and that they sold them green, shelled, at 2 cents, Mexican, per catty. At this price and yield his return would be $15.48, gold, per acre. If there was need of nitrogen and organic matter in the soil the vines would be pulled

Fig. 77. – In the lower section, along the path, basketfuls of canal mud had been applied in two rows at the rate of more than 100 tons per acre. In the middle section, workmen just beyond the extreme right were removing mud from such a reservoir as is seen in Fig. 48. The upper section shows three men distributing canal mud between the rows of a field of windsor beans.

green, after picking the beans, and composted with the wet mud. If not so needed, the dried stems would be tied in bundles and sold as fuel, or used as fuel at home and the ashes returned to the fields. The windsor beans are thus an early crop grown for fertilizer and fuel as well as for food.

This farmer was paying his labourers 100 cash per day and providing their meals, which he estimated worth 200 cash more, making 12 cents, gold, for a ten-hour day. Judging from what we saw and from the amount of mud carried per load, we estimated the men would distribute not less than eighty-four loads of 80 pounds each per day, an average distance of 500 feet, making the cost 3·57 cents, gold, per ton for distribution.

The lower section of Fig. 77 shows mud being used on a narrow strip bordering the path along which we walked. The amount shown in the illustration had been brought more than 400 feet by one man before 10 a.m. on the morning the photograph was taken. He was getting it from the bottom of a canal 10 feet deep, laid bare by the outgoing tide. Already he had brought more than a ton to his field.

The carrying baskets used for this work were in the form of huge dustpans suspended from the carrying poles by two cords attached to the side rims, and steadied by the hand grasping a handle provided in the back for this purpose as well as for emptying the baskets by tipping. With this construction the earth was readily raked upon the basket, and very easily emptied from it. No arrangement could be more expeditious or inexpensive for this man with his small holding. In such simple manner has nearly all the earth been moved in digging the miles of canal and in building the long sea walls. In Shanghai we saw the mud which is carried through the storm sewers into Soochow creek being removed in the same manner during the intervals when the tide was out.

In still another field (Fig. 78) canal mud had been applied at a rate exceeding 70 tons per acre, and we were told that such dressings would be repeated as often as every two years if other and cheaper fertilizers could be obtained. In the lower portion of Fig. 78 may be seen the section of canal from which this mud was taken up the three earthen stairways built of the mud itself. Many such lines of stairway were seen during our trips along the canals. To facilitate collecting the mud from the shallow canals temporary dams are thrown across them at two places and the water between the dams scooped or pumped out, laying the botton bare. The earth of the large grave mound seen across a canal in

Fig. 78. – Section of field covered with piles of canal mud recently applied at the rate of more than 70 tons per acre; taken out of the canal up the three flights of earth steps shown in the lower part of the figure.

the centre background of the upper portion of the engraving had been collected in this manner.

In the Chekiang province canal mud is extensively used in the mulberry orchards as a surface dressing. We have referred to this practice in southern China, and Fig. 79 is a view taken south of Kashing early in April. The boat anchored in front of the mulberry orchard is the home of a family coming from a distance, seeking employment during the season for picking mulberry leaves to feed silkworms. We were much surprised, on looking back at the boat after closing the camera, to see the head of the family standing erect in the centre, having shoved back a section of the matting roof.

The dressing of mud applied to this field formed a loose layer more than two inches deep, and when compacted by the rains which would follow would add not less than a full inch of soil over the entire orchard. The weight per acre could not be less than 120 tons.

Another equally, or even more, laborious practice followed by the Chinese farmers in this province is the periodic exchange of soil between mulberry orchards and the rice fields, their experience being that soil long used in the mulberry orchards improves the rice, while soil from the rice fields is very helpful when applied to the mulberry orchards. We saw many instances, when travelling by boat-train between Shanghai, Kashing and Hangchow, of soil being carried from rice fields and either stacked on the banks or dropped into the canal. Such soil was oftenest taken from narrow trenches leading through the fields. It is our judgment that the soil thrown into the canals undergoes important changes – perhaps through the absorption of soluble plant-food substances, such as lime, phosphoric acid and potash, withdrawn from the water, or through some growth or fermentation – which, in the judgment of the farmer, make the large labour involved in this procedure worth while. The stacking of soil along the banks was probably in preparation for its removal by boat to some of the mulberry orchards.

It is clearly recognized by the farmers that mud collected from those sections of the canal leading through country villages, such as that seen in Fig. 8, is both inherently more fertile and in better physical condition than that collected in the open country. They attribute this difference to the effect of the village washing in the canal, where soap is extensively used. The storm waters of the city doubtless carry some fertilizing material also, although sewage, as such, never finds its way into the canals. The washing

FIG. 79. – Mulberry orchard to which a heavy dressing of canal mud had been applied. A family of mulberry-leaf pickers were living in the boat anchored in the canal.

would be very likely to have a decided flocculating effect and so render the material more friable when applied to the field.

One very important advantage which comes to the fields when heavily dressed with such mud is the addition of lime which has become incorporated with the silts through their flocculation and precipitation, as well as that which is added in the form of snail shells abounding in the canals. The amount of the latter may be realized from the large number of shells contained in the mud recently thrown out, as seen in the upper section of Fig. 80, where the pebbly appearance of the surface is due to this cause. In the lower section of the same illustration the white spots are snail shells exposed in the soil of a recently spaded field. The shells are by no means as numerous generally as here seen, but they are sufficient to maintain the supply of lime.

Several species of these snails are collected in quantities and used as food. Piles containing bushels of the empty shells were seen along the canals outside the villages. The snails are cooked in the shell and often sold by measure to be eaten from the hand, as we buy roasted peanuts or pop-corn. When a purchase is made the vender clips the spiral point from each shell with a pair of small shears. This admits air and permits the snail to be readily removed by suction when the lips are applied to the shell. In the canals there are also large numbers of freshwater eels, shrimps and crabs as well as fish, all of which are collected and used for human food. It is common, when walking through the canal country, to come upon groups of gleaners busy in the bottoms of the shallow agricultural canals, gathering anything which may serve as food, even including small bulbs or the fleshy roots of edible aquatic plants. To facilitate the collection of such food materials, sections of the canal are often drained in the manner already described, so that gleaning may be done by hand, wading in the mud. Families living in house-boats make a business of fishing for shrimp. They trail behind the house-boat one or two other boats carrying hundreds of shrimp-traps, cleverly constructed in such a manner that when they are trailed along the bottom the shrimps dart into holes in the trap, mistaking them for safe hiding-places.

Travel between Shanghai and Hangchow at times is very heavy. Trains of six or more house-boats, each towed by a steam launch, are run by various companies and are daily crowded with passengers. Our train left Shanghai at 4.30 p.m., and reached Hangchow at 5.30 p.m. the following day, covering a distance of more than 117 miles. We paid $5.16, gold, for the exclusive use of a first-

FIG. 80. – The recently removed canal mud, in the upper section of the illustration, is heavily charged with large snail shells. The lower section shows the shells in the soil of a recently spaded field.

cabin, five-berth stateroom for myself and interpreter. It occupied the full width of the boat, except about 14 inches of footway, and could be entered from either side down a flight of five steps. The berths were flat, naked wooden shelves 30 inches wide, separated by a partition headboard 6 inches high and without railing in front. Each traveller provided his own bedding. A small table upon which meals were served, a mirror on one side and a lamp on the other, set in an opening in the partition, permitting it to serve two cabins, completed the furnishings. The roof of the cabins was covered with an awning and divided crosswise into two lines of berths, each 30 inches wide, by board partitions 6 inches high. In these sections passengers spread their beds, sleeping with their heads separated only by a headboard 6 inches high. The awning was just high enough to permit passengers to sit erect. Ventilation was ample, but privacy was nil.

Meals were served to each passenger wherever he might be. Dinner consisted of hot steamed rice brought in heavy porcelain bowls inside a covered hot wooden case. With the rice were tiny dishes of green clover, nicely cooked and seasoned; of cooked bean curd served with shredded bamboo sprouts; of tiny pork strips with bean curd; of small bits of liver with bamboo sprouts; and of greens. Hot water was provided for tea. There was no table-linen, and everything but the tea had to be negotiated with chop-sticks or, these failing, with the fingers. When the meal was finished the table was cleared and water, hot if desired, was brought for your hand-basin, which with tea, tea-cup and bedding, constitute part of the traveller's outfit. At frequent intervals, up to 10 p.m., a crier walked about the deck with hot water for those who might desire an extra cup of tea, and again in the early morning.

At this season of the year Chinese incubators were being run to their full capacity and it was our good fortune to visit one of these. The art of incubation is very old and very extensively practised in China. An interior view of one of these establishments is shown in Fig. 81. Here the family were hatching the eggs of hens, ducks and geese, purchasing the eggs and selling the young as hatched. As in the case of so many trades in China, this family was the last generation of a long line whose lives had been spent in the same work. We entered through their shop opening on the street of the narrow village seen in Fig. 8. In the shop the eggs were purchased and the chicks were sold, this work being in charge of the women of the family. It was in the extreme rear of the home that the incubators were installed, each having a capacity of 1,200 hens' eggs. Four of these may be seen in the illustration and one of the baskets which, when two-thirds filled with eggs, is set inside of each incubator.

Each incubator consists of a large earthenware jar with a door cut in one side through which live charcoal may be introduced and the fire partly smothered under a layer of ashes. The jar is thoroughly insulated, cased in basket-work and provided with a cover, as seen in the illustration. Inside the outer jar rests a second of nearly the same size, as one tea-cup may in another. Into this is lowered the large basket with its 600 hens' eggs, 400

Fig. 81. – Four Chinese incubators in a room where there are thirty, each having a capacity of 1,200 hens' eggs.

ducks' eggs or 175 geese' eggs, as the case may be. Thirty of these incubators were arranged in two parallel rows of fifteen each. Immediately above each row, and utilizing the warmth of the air rising from them, was a continuous line of finishing hatchers and brooders in the form of woven shallow trays with sides warmly padded with cotton and with the tops covered with sets of quilts of different thickness. It is in these brooder trays that the chicks emerge from the shells, and remain till ready to be taken to the shop and sold.

After a basket of hens' eggs has been incubated four days the eggs are examined, and those which are infertile are removed before they have been rendered unsaleable. The infertile eggs go to the shop. Ducks' eggs are similarly examined after two days, and again after five days' incubation; and geese' eggs after six days and again after fourteen days. Through these precautions practically all loss from infertile eggs is avoided and from 95 to 98 per cent of the fertile eggs are hatched, the infertile eggs ranging from 5 to 25 per cent.

After the fourth day in the incubator all eggs are turned five times in twenty-four hours. Hens' eggs are kept in the lower incubator eleven days, ducks' eggs thirteen days, and geese' eggs sixteen days, after which they are transferred to the trays. Throughout the incubation period the most careful watch and control is kept over the temperature. Different temperatures are maintained during different stages of the incubation. No thermometer is used, but the operator raises the lid or quilt, removes an egg and presses the large end into his eye-socket. In this way a large contact is made where the skin is sensitive, nearly constant in temperature, but little below blood heat and from which the air is excluded for the time. Long practice permits them thus to judge small differences of temperature expeditiously and with great accuracy. The men sleep in the room and some one is on duty continuously, making the rounds of the incubators and brooders, examining and regulating each according to its individual needs, through the management of the doors or the shifting of the quilts over the eggs in the brooder trays. In the finishing trays the eggs form rather more than one continuous layer, but the second layer does not cover more than a fifth or a quarter of the area. Hens' eggs are in these trays ten days, ducks' and geese' eggs fourteen days.

After the chickens have been hatched sufficiently long to require feeding they are ready for market, and are then sorted according to sex and placed in separate shallow woven trays 30 inches in diameter. The sorting is done rapidly and accurately through the sense of touch, the operator recognizing the sex by gently pinching the anus. Four trays of young chickens were in the shop fronting on the street as we entered and several women were making purchases, taking five to a dozen each. I was informed that nearly every family in the cities and in the country villages raise a few, but only a few, chickens, and it is a common sight to see grown chickens walking about the narrow streets, in and out of the open shops, dodging the feet of the occupants and passers-by. At the time of our visit this family was paying at the rate of 10 cents, Mexican, for nine hens' and eight ducks' eggs, and were selling their largest strong chickens at 3 cents each. These figures, translated into our currency, make the purchase price for eggs nearly 48 cents, and the selling price for the young chicks $1·29 per hundred, or thirteen eggs for 6 cents and seven chickens for 9 cents.

It is difficult even to conceive, not to say measure, the vast

FIG. 82. – Boat-load of 150 baskets of eggs on Soochow creek, Shanghai.

import of this solution of how to maintain, in the millions of homes, a constantly accessible supply of absolutely fresh and thoroughly sanitary animal food in the form of meat and eggs. The great density of population in these countries makes the problem of supplying eggs to the people very different from that in the United States. Our 250,600,000 fowls in 1900 were at the rate of three to each person, but in Japan, with her 16,500,000 fowls, there was in 1906 but one for every three people. The number per square mile of cultivated land, however, was 825, while in the United States, in 1900, the number of fowls per square mile of improved farm land was but 387. To give to Japan three fowls to each person there would need to be an average of about nine to each acre of her cultivated land, whereas in the United States there were in 1900 nearly two acres of improved farm land for each fowl. We have no statistics regarding the number of fowls in China or the number of eggs produced, but the total is very large and she exports to Japan. The large boat-load of eggs seen in Fig. 82 had just arrived from the country, coming into Shanghai in one of her canals.

FIG. 83. – Eight bearers moving a pile of winter compost to the recently excavated pit in the field seen in Fig. 84. The boat-load in the foreground is a mixture of manure and ashes just arrived from the home village.

Besides applying canal mud directly to the fields in the ways described there are other very extensive practices of composting it with organic matter of one kind or another and of then using the compost on the fields. The next three illustrations show some of the different stages in the process as well as the tremendous labour of body and amount of forethought required. In Fig. 83 eight bearers may be seen moving winter compost to a recently excavated pit in an adjoining field, shown in Fig. 84. Four months before men had brought waste from the stables of

Fig. 84. – Compost pit adjacent to a field of clover, being filled from the pile of winter compost seen on the bank of the canal in Fig. 83.

Shanghai, a distance of 15 miles by water, deposited it upon the canal bank between layers of thin mud dipped from the canal, and left it to ferment. The eight men were removing this compost to the pit seen in Fig. 85, then nearly filled. Near by in the same field was a second pit, shown in Fig. 85, excavated 3 feet deep and rimmed about with the earth removed, making it 2 feet deeper.

After the pits had been filled the clover which was in blossom beyond the pits would be cut and stacked upon them to a height of 5 to 8 feet, and this also saturated, layer by layer, with mud

FIG. 85. – Recently excavated pit for receiving winter compost seen in Fig. 83, and upon which the clover beyond the pit will be cut and composted as a fertilizer for a crop of rice.

brought from the canal. It would then be allowed to ferment twenty to thirty days, until the juices set free had been absorbed by the winter compost beneath, and until the time had arrived for fitting the ground for the next crop. This organic matter, fermented with the canal mud, would then be distributed by the men over the field, carried for the third time on their shoulders, notwithstanding its weight amounted to many tons.

The manure had been collected, loaded and carried 15 miles by water; it had been unloaded upon the bank and saturated with canal mud; the field had been fitted for clover the previous autumn and seeded; the pits had been dug in the fields; the winter compost had been carried and placed in the pits; the clover was to be cut, carried by the men on their shoulders, stacked layer by layer and saturated with mud dipped from the canal; the whole would later be distributed over the field, and finally the earth removed from the pits would be returned to them, that the service of no ground upon which a crop might grow should be lost.

Such are the tasks to which Chinese farmers hold themselves, because they are convinced desired results will follow, because their holdings are so small and their families so large. These practices are so extensive in China and so fundamental in the part they play in the maintenance of high productive power in their soils

Fig. 86. – Providing for the building of a mud-and-clover compost stack.

that we made special effort to follow them through different phases. In Fig. 86 we saw the preparation being made to build one of the clover compost stacks saturated with canal mud. On the left the thin mud had been dipped from the canal. Wayfarers in the centre were crossing the foot-bridge of the country by-way. Beyond rose the conical thatch to shelter the water buffalo when pumping the water for irrigating the rice crop which was to be fed with the plant food now in preparation. On the right were two large piles of green clover freshly cut. A woman of the family at one of them was spreading it to receive the mud, while the men-folk were coming from the field with more clover on their carrying poles. We came upon this scene just before the dinner hour and, after the workers had left, another photograph was

Fig. 87. – Clover compost stack in the building.

taken at closer range from a different side, giving the view seen in Fig. 87. As the mud had been removed some days and become too stiff to spread, water was being brought from the canal in the pails at the right for reducing its consistency to that of a thin porridge, permitting it more completely to smear and saturate the clover. The stack grew, layer by layer, each saturated with the mud, tramped solid with the bare feet. Provision had been made here for building four other stacks.

Farther along we came upon the scene in Fig. 88, where the

Fig. 88. – The young man is loading his boat with canal mud, using the long-handled clam-shell dredge; which he can open and close at will.

building of the stack of compost and the gathering of the mud from the canal were simultaneous. On one side of the canal the son, using a clam-shell shaped dipper made of basket-work, which could be opened and shut with a pair of bamboo handles, had nearly filled the middle section of his boat with the thin ooze, while on the other side, against the stack which was building, the mother was emptying a similar boat, with a large dipper, also provided with a bamboo handle. The man on the stack is a good scale for judging its size.

We came next upon a finished stack on the bank of another

canal, shown in Fig. 89, where our umbrella was set to serve as a scale. This stack measured 10 by 10 feet on the ground, was 6 feet high and must have contained more than 20 tons of the green compost. At the same place, two other stacks had been started, each about 14 by 14 feet, and foundations were laid for six others, nine in all.

During twenty or more days this green nitrogenous organic matter is permitted to lie fermenting in contact with the fine soil particles of the ooze with which it had been charged. This is a

FIG. 89. – A completed compost stack.

remarkable practice in that it is a very old, intensive application of an important fundamental principle only recently understood and added to the science of agriculture, namely, the power of organic matter, decaying rapidly in contact with soil, to liberate from it soluble plant food. It would be a great mistake, therefore, to say that these laborious practices are the result of ignorance, of a lack of capacity for accurate thinking, or of power to grasp and utilize. If the agricultural lands of the United States are ever called upon to feed even 1,200 millions of people, a number proportionately less than one-half that being fed in Japan to-day, very different practices from those now followed will have been adopted. We can believe they will require less human bodily effort and be more efficient. But the knowledge which can make them so is not yet in the possession of our farmers, much less the conviction that plant feeding and more persistent and better-directed soil management are necessary to such yields as will then be required.

Later, just before the time for transplanting rice, we returned to the same district to observe the manner of applying the compost to the field, and Fig. 90 is prepared from photographs then taken, illustrating the activities of one family, as seen during the morning of May 28th. Their home was in a near-by village and their holding was divided into four nearly rectangular paddy fields, graded to water level, separated by raised rims, and having an area of nearly two acres. Three of these little fields are partly shown in the illustration, and the fourth in Fig. 143. In the background of the upper section of Fig. 90, and under the thatched shelter, was a native Chinese cow, blindfolded and hitched to the power-wheel of a large wooden-chain pump, lifting water from the canal and flooding the field in the foreground, to soften the soil for ploughing. Riding on the power-wheel was a girl of some twelve years, another of seven and a baby. They were there for entertainment and to see that the cow kept at work. The ground had been sufficiently softened and the father had begun ploughing, the cow sinking to her knees as she walked. In the same paddy field, but shown in the section below, a boy was spreading the clover compost with his hands, taking care that it was finely divided and evenly scattered. He had been once around before the ploughing began. The compost had been brought from a stack by the side of a canal, and two other men were busy, still bringing the material to one of the other paddy fields, one of whom, with his baskets on the carrying pole, appears in the third section. Between these two paddy fields was the one seen at the bottom of the illustration, which had matured a crop of rape that had been pulled and was lying in swaths ready to be moved. Two other men were busy here, gathering the rape into large bundles and carrying it to the village home, where the women were threshing out the seed, taking care not to break the stems, which, after threshing, were tied into bundles for fuel. The seed would be ground and from it an oil expressed, while the cake would be used as a fertilizer.

This crop of rape is remarkable for the way it fits into the

FIG. 90. – The activities of a family, fertilizing and fitting paddy fields for rice.

economies of these people. It is a near relative of mustard and cabbage and it grows rapidly during the cooler portions of the season, as the spring crop ripens before the planting of rice and cotton. Its young shoots and leaves are succulent, nutritious, readily digested and extensively used as human food, either boiled and eaten fresh, or salted for winter use, to be served with rice. The mature stems, being woody, make good fuel, and it bears a heavy crop of seed, rich in oil, extensively used for lighting and cooking, while the rape-seed cake is highly prized as a manure and extensively used. In the early spring the country is luxuriantly green with the large acreage of rape, later changing to a sea of most brilliant yellow and finally to an ashy grey when the leaves fall and the stems and pods ripen.

Like the dairy cow, rape produces a fat, in the ratio of about 40 pounds of oil to 100 pounds of seed, which may be eaten, burned or sold without materially robbing the soil of its fertility, provided that the cake and the ashes from the stems are returned to the fields, for the carbon, hydrogen and oxygen of which the oil is almost wholly composed come from the atmosphere rather than from the soil.

IX

THE UTILIZATION OF WASTE

ONE of the most remarkable agricultural practices adopted by any civilized people is the centuries-long and wellnigh universal conservation of all human waste in China, Korea and Japan, and its utilization in the maintenance of soil fertility and in the production of food. To understand this evolution, it must be recognized that mineral fertilizers so extensively employed in modern Western agriculture, like the extensive use of mineral coal, had been a physical impossibility to all people alike until within very recent years. With this fact must be associated the very long unbroken life of Eastern nations and the vast numbers their farmers have been compelled to feed.

When we reflect upon the depleted fertility of our own older farm lands, comparatively few of which have seen a century's service, and the enormous quantity of mineral fertilizers which are being applied annually to them in order to secure paying yields, it is evident that the time has come when profound consideration should be given to the practices the Mongolian race has maintained through many centuries.

From the analyses of mixed human excreta made by Wolff in Europe and by Kellner in Japan, it appears that, as an average, these carry in every 2,000 pounds 12·7 pounds of nitrogen, 4 pounds of potassium and 1·7 pounds of phosphorus. On this basis and that of Carpenter, who estimates the average amount of excreta per day for the adult at 40 ounces, the average annual production per million of adult population is 5,794,300 pounds of nitrogen, 1,825,000 pounds of potassium, and 775,600 pounds of phosphorus carried in 456,250 tons of excreta. The figures which Hall cites in *Fertilizers and Manures* would make these amounts 7,940,000 pounds of nitrogen, 3,070,500 pounds of potassium, and 1,965,600 pounds of phosphorus, but the figures he takes and calls high averages give 12,000,000 of nitrogen, 4,151,000 pounds of potassium, and 3,057,600 pounds of phosphorus.

In 1908 the International Concessions of the city of Shanghai sold to one Chinese contractor for $31,000, gold, the privilege of collecting 78,000 tons of human waste, and of removing it to the country for sale to farmers. The flotilla of boats seen in Fig. 91 is one of several engaged daily in Shanghai throughout the year in this service.

Dr. Kawaguchi, of the National Department of Agriculture and Commerce, taking his data from their records, informed us that the human manure saved and applied to the fields of Japan in 1908 amounted to 23,850,295 tons, which is an average of 1·75 tons per acre of their 21,321 square miles of cultivated land in the four main islands.

Fig. 91. – A flotilla of manure boats on Soochow creek, collecting human wastes in the city of Shanghai, for removal to cultivated fields.

On the basis of the data of Wolff, Kellner and Carpenter, or of Hall, the people of the United States and of Europe are pouring into the sea, lakes or rivers, and into the underground waters, from 5,791,300 to 12,000,000 pounds of nitrogen, 1,881,900 to 4,151,000 pounds of potassium, and 777,200 to 3,057,600 pounds of phosphorus per million of adult population annually, and this waste we esteem one of the great achievements of our civilization. In the Far East, for more than thirty centuries, these enormous wastes have been religiously saved, and to-day the 400 millions of adult population send back to their fields annually 150,000 tons of phosphorus, 376,000 tons of potassium, and 1,158,000 tons of nitrogen comprised in a gross weight exceeding 182,000,000 tons. They are gathered from every home, alike in country villages and in

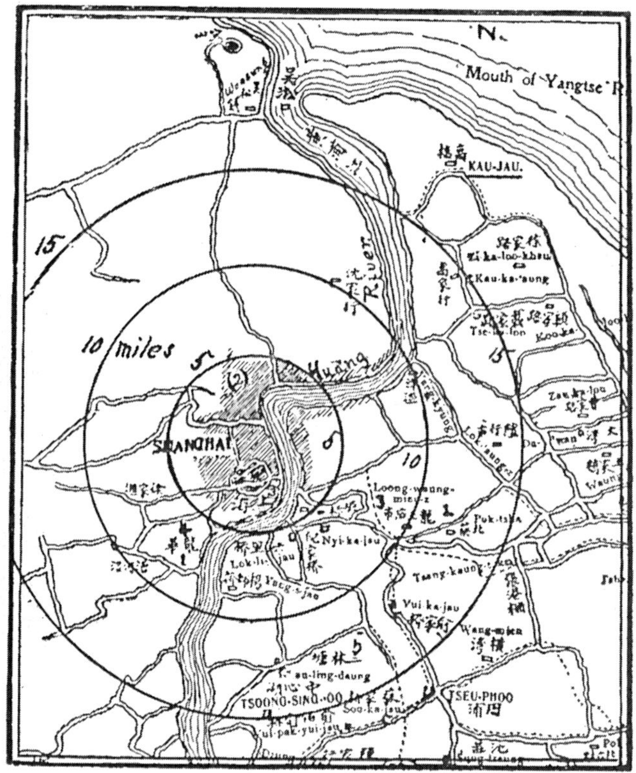

FIG. 92. – Map of country surrounding Shanghai, showing a few of the many canals on which the waste of the city is conveyed by boat to the farms.

great cities like Hankow-Wuchang-Hanyang with their 1,770,000 people swarming on a land area delimited by a radius of 4 miles.

Man is the most extravagant accelerator of waste the world has ever endured. His withering blight has fallen upon every living thing within his reach, himself not excepted; and his besom of destruction in the uncontrolled hands of a generation has swept into the sea soil-fertility which only centuries of life could accumulate – fertility which is the substratum of all that is living. It must be recognized that the phosphate deposits which we are beginning to return to our fields are but measures of fertility lost from older soils, and indices of processes still in progress. The rivers of North America are estimated to carry to the sea more than 500 tons of phosphorus with each cubic mile of water. To such loss modern civilization is adding that of hydraulic sewage disposal, through which the waste of 500 millions of people might be more than 194,300 tons of phosphorus annually, a waste which could not be replaced by 1,295,000 tons of rock phosphate, 75 per cent pure. The Mongolian races, with a population now approaching the figure named; occupying an area little more than one-half that of the United States, tilling less than 800,000 square miles of land, and much of this during twenty, thirty or perhaps forty centuries; unable to avail themselves of mineral fertilizers, could not tolerate such waste and survive. Compelled to solve the problem of avoiding such wastes, and exercising the faculty which is characteristic of the race, they 'cast down their buckets where they were,' as

[1] A ship lost at sea for many days suddenly sighted a friendly vessel. From the mast of the unfortunate vessel was seen a signal, 'Water, water; we die of thirst!' The answer from the friendly vessel at once came back, 'Cast down your bucket where you are.' A second time the signal, 'Water, water; Send us water!' ran up from the distressed vessel, and was answered, 'Cast down your bucket where you are.' And a third and fourth signal for water was answered, 'Cast down your bucket where you are.' The captain of the distressed vessel, at last heeding the injunction, cast down his bucket, and it came up full of fresh sparkling water from the mouth of the Amazon river.

Not even in great cities like Canton, built in the meshes of tide-swept rivers and canals; like Hankow on the banks of one of the largest rivers in the world; nor yet in modern Shanghai, Yokohama or Tokio is such waste permitted. To them such a practice would

[1] Booker T. Washington, Atlanta address.

have meant race suicide, and they have resisted the temptation so long that it has ceased to exist.

Dr. Arthur Stanley, Health Officer of the city of Shanghai, in his annual report for 1899, considering this subject as a municipal problem, wrote:

'Regarding the bearing on the sanitation of Shanghai of the relationship between Eastern and Western hygiene, it may be said, that if prolonged national life is indicative of sound sanitation, the Chinese are a race worthy of study by all who concern themselves with Public Health. Even without the returns of a Registrar-General it is evident that in China the birth-rate must very considerably exceed the death-rate, and have done so in an average way during the three or four thousand years that the Chinese nation has existed. Chinese hygiene, when compared with mediæval English, appears to advantage. The main problem of sanitation is to cleanse the dwelling day by day, and if this can be done at a profit so much the better. While the ultra-civilized Western elaborates destructors for burning garbage at a financial loss and turns sewage into the sea, the Chinaman uses both for manure. He wastes nothing while the sacred duty of agriculture is uppermost in his mind. And in reality recent bacterial work has shown that fæcal matter and house refuse are best destroyed by returning them to clean soil, where natural purification takes place. The question of destroying garbage can, I think, under present conditions in Shanghai, be answered in a decided negative. While to adopt the water-carriage system for sewage and turn it into the river, whence the water supply is derived, would be an act of sanitary suicide. It is best, therefore, to make use of what is good in Chinese hygiene, which demands respect, being, as it is, the product of an evolution extending from more than a thousand years before the Christian era.'

The storage of such waste in China is largely in stoneware receptacles, such as are seen in Fig. 93, which are hard-burned, glazed terra-cotta urns, having capacities ranging from 500 to 1,000 pounds. Japan more often uses sheltered cement-lined pits such as are seen in Fig. 94.

In the three countries the carrying to the fields is most often in some form of pail, as seen in Fig. 95, a pair of which are borne swinging from the carrying pole. In applying the liquid to the field or garden the long-handled dipper is used, seen in Fig. 96.

We are beginning to husband with some economy the waste from our domestic animals, but in this we do not approach that of China, Korea and Japan. People in China regularly search for and collect droppings along the country and caravan roads. Repeatedly, when walking through city streets, we observed such materials quickly and apparently eagerly gathered, to be carefully

FIG. 93. – Receptacles for human waste.

stored under conditions which ensure small loss from either leaching or unfavourable fermentation. In some mulberry orchards the earth had been carefully hoed back about the trunks of trees to a depth of 3 or 4 inches from a circle having a diameter of 6 to 8 feet, and upon these areas were placed the droppings of silkworms, the moulted skins, together with the bits of leaves and stem left after feeding. Some disposition of such waste must be made. They return at once to the orchard all but the silk produced from the leaves; unnecessary loss is thus avoided and the material enters at once the service of forcing the next crop of leaves.

FIG. 94. – Japanese sheltered cement-lined storage pits for liquid manure.

FIG. 95. – Six carrying pails such as are used in distributing liquid manure to the fields.

On the farm of Mrs. Wu, near Kashing, while studying the operation of two irrigation pumps driven by two cows, lifting water to flood her 25 acres of rice field preparatory to transplanting, we were surprised to observe that one of the duties of the lad who had charge of the animals was to use a six-quart wooden dipper with a bamboo handle 6 feet long to collect all excreta, before they fell upon the ground, and transfer them to a receptacle provided for the purpose. There came a flash of resentment that such a task was set for the lad, for we were only beginning to

FIG. 96. – Applying liquid manure from carrying pails, using the long-handled dipper.

realize to what lengths the practice of economy may go, but there was nothing irksome suggested in the boy's face. He performed the duty as a matter of course, and as we thought it through there was no reason why it should have been otherwise. In fact, the only right course was being taken. Conditions would have been worse if the collection had not been made. It made possible more rice. Character of substantial quality was building in the lad which meant thrift in the growing man and continued life for the nation.

Much intelligence and the highest skill are exhibited by these old-world farmers in the use of their wastes. In Fig. 97 is one of many examples which might be cited. The man walking down the row with his manure pails swinging from his shoulders informed us on his return that in his household there were twenty to be fed; that from this garden of half an acre of land he usually sold a product bringing in $100, Mexican — $172, gold. The crop was cucumbers in groups of two rows 30 inches apart and 24 inches between the groups. The plants were 8 to 10 inches apart in the row. He had just marketed the last of a crop of greens which occupied the space between the rows of cucumbers seen under the strong, durable, light and very readily removable trellises. On May 28th the vines were beginning to run, so not a minute had been lost in the change of crop. On the contrary, this man had added a month to his growing season by over-lapping his crops, and the trellises enabled him to feed more plants of this type than there was room for vines on the ground. With ingenuity and much labour he had made his half-acre for cucumbers equivalent to more than two. He had removed the vines entirely from the ground; had provided a travel space 2 feet wide, down which he was walking, and he had made it possible to work about the roots of every plant for the purpose of hoeing and feeding. Four acres of cucumbers handled by American field methods would not yield more than this man's one, and he grows besides two other crops the same season. The difference is not so much in activity of muscle as it is in alertness and efficiency of the grey matter of the brain. He sees and treats each plant individually, he loosens the ground so that his liquid manure drops immediately beneath the surface within reach of the active roots. If the rainfall has been scanty and the soil is dry he may use ten of water to two of night soil, not to supply water but to make certain sufficiently deep penetration. If the weather is rainy and the soil over wet, the food is applied more concentrated, not to lighten the burden but to avoid waste by leaching and over-saturation. While he is ever crowding growth he never overfeeds. Forethought, after-thought and the mind focused on the work in hand are characteristic of these people. We do not recall to have seen a man smoking while at work. They enjoy smoking, but prefer to do this also with the attention undivided and thus get more for their money.

On another date earlier in May we were walking in the fields

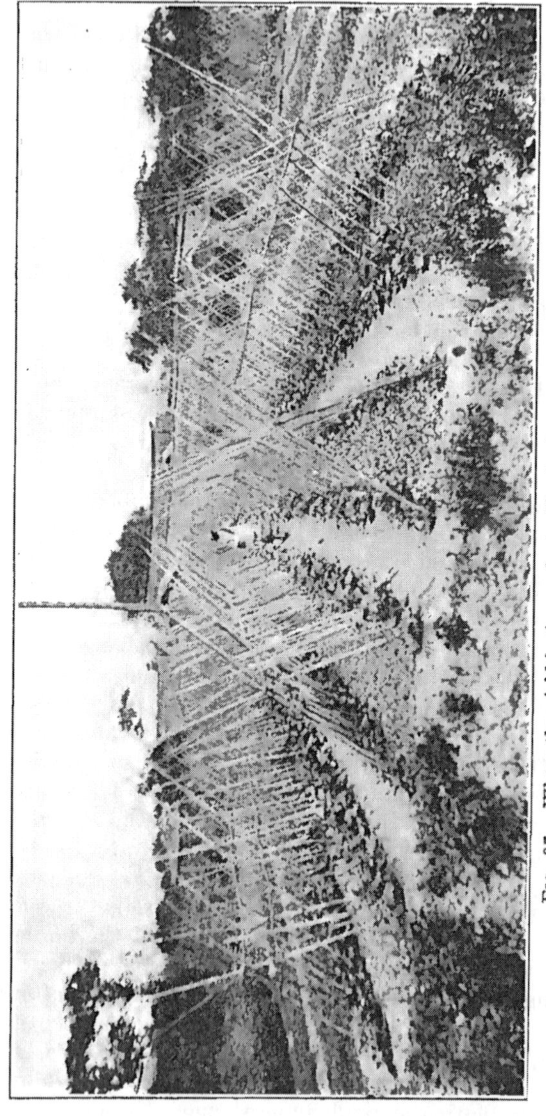

Fig. 97. – Where the yield is the product of brain, brawn and utilized waste.

without an interpreter. For half an hour we stood watching an old gardener fitting the soil with his spading hoe in the manner seen in Fig. 22, where the graves of his ancestors occupy a part of the land. Angle-worms were extremely numerous, as large round as an ordinary lead pencil, and, when not extended, two-thirds as long, decidedly greenish in colour. Nearly every stroke of the spade exposed two to five of these worms, but so far as we observed, and we watched the man closely, pulverizing the soil, he neither injured nor left uncovered a single worm. While he seemed to make no effort to avoid injuring them or to cover them with earth, and while we could not talk with him, we are convinced that his action was continually guarded against injuring the worms. They certainly were subsoiling his garden deeply and making possible a freer circulation of air far below the surface. Their great abundance proved a high content of organic matter present in the soil and, as the worms ate their way through it, passing the soil through their bodies, the yearly volume of work done by them was very great. In the fields flooded preparatory to fitting them for rice these worms are forced to the surface in enormous numbers, and large flocks of ducks are taken to such fields to feed upon them.

In another field a crop of barley was nearing maturity. An adjacent strip of land was to be fitted and planted. The leaning barley heads were in the way. Not one must be lost and every inch of ground must be put to use. The grain along the margin, for a breadth of 16 inches, had been gathered into handfuls and skilfully tied, each with an unpulled barley stem, without breaking the straw, thus permitting even the grains in that head to fill and be gathered with the rest, while the tying set all straws well aslant, out of the way, and permitted the last inch of naked ground to be fitted without injuring the grain.

In still another instance a man was growing Irish potatoes to market when yet small. He had enriched his soil; he would have applied water if the rains had not been timely and sufficient; and he had fed the plants. He had planted in rows only 12 to 14 inches apart with a hill every 8 inches in the row. The vines stood strong, straight, 14 inches high and as even as a trimmed hedge. The leaves and stems were turgid, the deepest green, and as prime and glossy as a prize steer. So close were the plants that there was leaf surface to intercept the sunshine falling on every square inch of the patch. There were no potato beetles and we saw no signs of injury, but the gardener was scanning the patch with the eye of a robin. He spied the slightest first drooping of leaves in a stem; went after the difficulty and brought and placed in our hand a cut-worm, a young tuber the size of a marble and a stem cut half off, which he was willing to sacrifice because of our evident interest.

But the two friends who had met were held apart by the confusion of tongues. Nothing is costing the world more, has made so many enemies, and has so much hindered the forming of friendships as the inability to fully understand. Hence the dove that brings world peace must fly on the wings of a common language. The bright star in the east is world commerce, rising on rapidly developing railway and steamship lines, heralded and directed by electric communication. With world commerce must come mutual confidence and friendship, requiring a full understanding and therefore a common tongue. Then world peace will be permanently assured. It is coming inevitably and faster than we think. Once this desired end is seriously sought, the carrying of three generations of children through the public schools where the world language is taught together with the mother tongue, and the passing of the parents and grandparents, would effect the change.

The important point regarding these Far Eastern peoples, to which attention should be directed, is that effective thinking, clear and strong, prevails among the farmers who have fed and are still feeding the dense populations from the products of their limited areas. This is further indicated in the universal and extensive use of plant ashes derived from fuel grown upon cultivated fields and upon the adjacent hill and mountain lands.

We were unable to secure exact data regarding the amount of fuel burned annually in these countries, and of ashes used as fertilizer, but a cord of dry oak wood weighs about 3,500 pounds, and the weight of fuel used in the home and in manufactures must exceed that of two cords per household. Japan has an average of 5·563 people per family. If we allow but 1,300 pounds of fuel *per capita*, Japan's consumption would be 31,200,000 tons. In view of the fact that a very large share of the fuel used in these countries is either agricultural plant stems, with an average ash content of 5 per cent., or the twigs and even leaves of trees, as in the case of pine-bough fuel, 4·5 per cent. of ash may be taken as a fair estimate. On this basis, and with a content of phosphorus equal to ·5 per cent, and of potassium equal to 5 per cent, the fuel ash for Japan would amount to 1,404,000 tons annually, carrying 7,020 tons of phosphorus and 70,200 tons of potassium, together with more than 400,000 tons of limestone, returned annually to less than 21,321 square miles of cultivated land.

In China, with her more than 400 millions of people, a similar rate of fuel consumption would make the phosphorus and potassium returned to her fields more than eight times the amounts computed for Japan. On the basis of these statements, Japan's annual saving of phosphorus from the waste of her fuel would be equivalent to more than 46,800 tons of rock phosphate having a purity of 75 per cent, or in the neighbourhood of 7 pounds per acre. If this amount, even with the potash and limestone added, appears like a trifling addition of fertility, it is important to remember that, even if this is so, these people have felt compelled to make the saving.

In the matter of returning soluble potassium to the cultivated fields Japan would be applying with her ashes the equivalent of no less than 156,600 tons of pure potassium sulphate, equal to 23 pounds per acre; while the lime carbonate so applied annually would be some 62 pounds per acre.

In addition to the forest lands, which have long been made to contribute plant food to the cultivated fields through fuel ashes, there are large areas which contribute green manure and compost material. These are chiefly hill lands, aggregating some 20 per cent of the cultivated fields, which bear mostly herbaceous growth. Some 2,552,741 acres of these lands may be cut over three times each season, yielding, in 1903, an average of 7,980 pounds per acre. The first cutting of this hill herbage is mainly used on the rice fields as green manure, trampled into the mud between the rows after the manner seen in Fig. 98. The man had been with basket and sickle to gather green herbage wherever he could and had brought it to his rice field. The day was in July and extremely sultry. We came upon him wading in the water half-way to his knees, carefully laying the herbage between alternate rows of rice, one handful in each place, with tips overlapping. This done, he took the attitude seen in the illustration and, gathering the materials into a compact bunch, pressed it beneath the surface with his foot. The two hands smoothed the soft mud over the grass and righted the disturbed spears of rice in the two adjacent hills. Thus, foot following foot, one bare length ahead, the succeeding bunches of herbage were submerged until the last had been reached.

He was renting the land at 40 kan of rice per tan, and his usual

FIG. 98. – Japanese farmer trampling green herbage for fertilizer into the water and mud between rows of rice.

yield was 80 kan. This is 44 bushels of 60 pounds per acre. In unfavourable seasons his yield might be less, but still his rent would be 40 kan per tan, unless it was clear that he had done all that could reasonably be expected of him in securing the crop.

The second and third cuttings of herbage from the *genya* lands in Japan are used for the preparation of compost applied on the dry-land fields in the autumn or in the spring of the following season. Some of these lands are pastured, but approximately 10,185,500 tons of green herbage grown and gathered from the hills contributes much of its organic matter and all of its ash to enrich the cultivated fields. Such wild growth areas in Japan are the commons of the near-by villages, to which the people are freely admitted for the purpose of cutting the herbage. A fixed time may be set for cutting and a limit placed upon the amount which may be carried away, which is done in the manner seen in Fig. 99. It is well recognized by the people that this constant cutting and

FIG. 99. – Father and children returning from *genya* lands with herbage for use as green manure or for making compost. The daughter carries the tea-kettle to supply their safe sanitary drink.

removal of growth from the hill lands, with no return, depletes the soils and reduces the amount of green herbage they are able to secure.

Through the kindness of Dr. Daikuhara, of the Imperial Agricultural Experiment Station at Tokio, we are able to give the average composition of the green leaves and young stems of five of the most common wild species of plants cut for green manure in June.

In each 1,000 pounds the amount of water is 562·18 pounds; of organic matter, 382·68 pounds; of ash, 55·14 pounds; nitrogen, 4·78 pounds; potassium, 2·407 pounds; and phosphorus, ·34 pound. On the basis of this composition and an aggregate yield of 10,185,500 tons, there would be annually applied to the cultivated fields 3,463 tons of phosphorus and 24,516 tons of potassium derived from the *genya* lands.

In addition to this the run-off from both the mountain and the *genya* lands is largely used upon the rice fields, more than 16 inches of water being applied annually to them in some prefectures. If such waters have the composition of river waters in North America, 12 inches of water applied to the rice fields of the three main islands would contribute no less than 1,200 tons of phosphorus and 19,000 tons of potassium annually.

Dr. Kawaguchi, of the National Department of Agriculture and Commerce, informed us that in 1908 Japanese farmers prepared and applied to their fields 22,812,787 tons of compost manufactured from the wastes of cattle, horses, swine and poultry, combined with herbage, straw and other similar wastes and with soil, sod or mud from ditches and canals. The amount of this compost is sufficient to apply 1·78 tons per acre of cultivated land of the southern three main islands.

From data obtained at the Nara Experiment Station, the composition of compost as there prepared shows it to contain, in each 2,000 pounds, 550 pounds of organic matter; 15·6 pounds of nitrogen; 8·3 pounds of potassium; and 5·24 pounds of phosphorus. On this basis 22,800,000 tons of compost will carry 59,700 tons of phosphorus and 94,600 tons of potassium. The construction of compost houses is illustrated in Fig. 100, reproduced from a large circular sent to farmers from the Nara Experiment Station, and an exterior of one at the Nara Station is given in Fig. 101.

This compost house is designed to serve 2½ acres. Its floor is 12 by 18 feet, rendered water-tight by a mixture of clay, lime and sand. The walls are of earth 1 foot thick, and the roof is thatched with straw. Its capacity is 16 to 20 tons, having a cash value of 60 yen, or $30. In preparing the stack, materials are brought daily and spread over one side of the compost floor until the pile has attained a height of 5 feet. After 1 foot in depth has been laid and firmed, 1·2 inches of soil or mud is spread over the surface and the process repeated until full height has been attained. Water is added sufficient to keep the whole saturated and to maintain the temperature below that of the body. After the compost stacks have been completed they are permitted to stand five weeks in summer, seven weeks in winter, when they are forked over and transferred to the opposite side of the house.

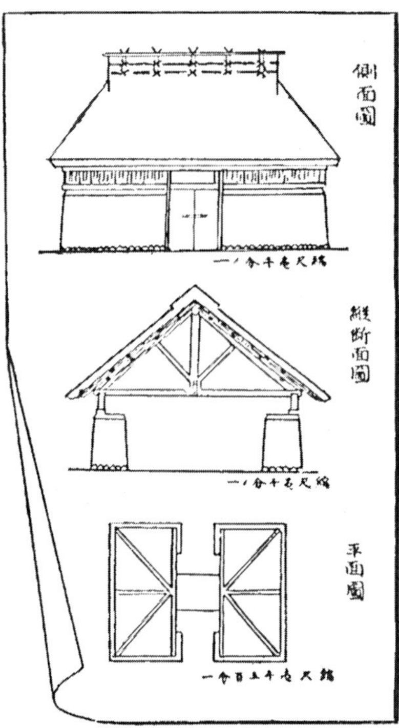

FIG. 100. – Section of chart issued by the Nara Experiment Station, illustrating construction of compost house; upper section shows elevation; middle portion is a cross section, and the lower shows floor plan.

If we state in round numbers the total nitrogen, phosphorus and potassium thus far enumerated which Japanese farmers apply or return annually to their 20,000 or 21,000 square miles of cultivated fields, the case stands 385,214 tons of nitrogen, 91,656 tons of phosphorus, and 255,778 tons of potassium. These values are only approximations and do not include the large volume and variety of fertilizers prepared from fish, which have long been used. Neither do they include the very large amount of nitrogen derived directly from the atmosphere through their long, extensive and persistent cultivation of soy beans and other legumes. Indeed, from 1903 to 1906 the average area of paddy fields upon which was grown a second crop of green manure in the form of some legume was 6·8 per cent. of the total area of such fields, aggregating 11,000 square miles. In 1906 over 18 per cent of

FIG. 101. – Exterior view of compost house at Nara Experiment Station.

the upland fields, aggregating between 9,000 and 10,000 square miles, also produced some leguminous crop.

While the values which have been given above, expressing the sum total of nitrogen, phosphorus and potassium applied annually to the cultivated fields of Japan, may be somewhat too high for some of the sources named, there is little doubt that Japanese farmers apply to their fields more of these three plant-food elements annually than has been computed. The amounts which have been given are sufficient to provide annually, for each acre of the 21,321 square miles of cultivated land, an application of not less than 56 pounds of nitrogen, 13 pounds of phosphorus and 37 pounds of potassium. Or, if we omit the large northern island of Hokkaido, still new in its agriculture and lacking the intensive practices of the older farm land, the quantities are sufficient for a mean application of 60, 14 and 40 pounds respectively of nitrogen, phosphorus and potassium per acre; and yet the maturing of 1,000 pounds of wheat crop, covering grain and straw as water-free substance, removes from the soil but 13·9 pounds of nitrogen, 2·3 pounds of phosphorus and 8·4 pounds of potassium, from which it may be computed that the 60 pounds of nitrogen added is sufficient for a crop yielding 31 bushels of wheat; the phosphorus is sufficient for a crop of 44 bushels, and the potassium for a crop of 355 bushels per acre.

Dr. Hopkins, in his recent valuable work on *Soil Fertility and Permanent Agriculture*, gives, on page 154, a table from which we abstract the following data:

APPROXIMATE AMOUNTS OF NITROGEN, PHOSPHORUS AND POTASSIUM REMOVABLE PER ACRE ANNUALLY BY

	Nitrogen, pounds.	Phosphorus, pounds.	Potassium, pounds.
100 bush. crop of corn	148	23	71
100 bush. crop of oats	97	16	68
50 bush. crop of wheat	96	16	58
25 bush. crop of soy beans	159	21	73
100 bush. crop of rice	155	18	95
3 ton crop of timothy hay	72	9	71
4 ton crop of clover hay	160	20	120
3 ton crop of cow pea hay	130	14	98
8 ton crop of alfalfa hay	400	36	192
7,000 lb. crop of cotton	168	29·4	82
400 bush. crop of potatoes	84	17·3	120
20 ton crop of sugar beets	100	18	157
Annually applied in Japan, more than	60	14	40

We have inserted in this table, for comparison, the crop of rice, and have increased the crop of potatoes from 300 bushels to 400 bushels per acre, because such a yield, like all of those named, is quite practicable under good management and favourable seasons, notwithstanding the fact that much smaller yields are generally attained through lack of sufficient plant food or water. From this table, assuming that a crop of matured grain contains 11 per cent of water and the straw 15 per cent, while potatoes contain 79 per cent and beets 87 per cent, the amounts of the three plant-food elements removable annually by 1,000 pounds of crop have been calculated and stated in the next table.

APPROXIMATE AMOUNTS OF NITROGEN, PHOSPHORUS AND POTASSIUM REMOVABLE ANNUALLY PER 1,000 POUNDS OF DRY CROP SUBSTANCE

	Nitrogen, pounds.	Phosphorus, pounds.	Potassium, pounds.
Cereals.			
Wheat	13·873	2·312	8·382
Oats	13·666	2·254	9·580
Corn	13·719	2·149	6·676
Legumes.			
Soy beans	30·807	4·070	14·147
Cow peas	25·490	2·745	19·216
Clover	23·529	2·941	17·647
Alfalfa	29·411	2·647	14·118
Roots.			
Beets	19·213	3·462	30·192
Potatoes	15·556	3·210	22·222
Grass.			
Timothy	14·117	1·765	13·922
Rice	9·949	1·129	6·089

From the amounts of nitrogen, phosphorus and potassium applied annually to the cultivated fields of Japan and from the data in these two tables, it may be readily seen that these people are now, and probably long have been, applying quite as much of these three plant-food elements to their fields with each planting as are removed from the crop, and if this is true in Japan it must also be true in China. Moreover, there is nothing in American agricultural practice which indicates that we shall not ultimately be compelled to do likewise.

X

IN THE SHANTUNG PROVINCE

On May 15th we left Shanghai by one of the coastwise steamers for Tsingtao, some 300 miles farther north, in the Shantung province, our object being to keep in touch with methods of tillage and fertilization, corresponding phases of which would occur there later in the season.

The Shantung province is in the latitude of North Carolina and Kentucky, or lies between that of San Francisco and Los Angeles. It has an area of nearly 56,000 squares miles, about that of Wisconsin. Less than one-half of this area is cultivated land, yet it is at the present time supporting a population exceeding 38,000,000 of people. New York state has to-day less than 10 millions and more than half of these are in New York city.

It was in this province that Confucius was born, 2,461 years ago, and that Mencius, his disciple, lived. Here, too, 1,700 years before Confucius' time, after one of the great floods of the Yellow River, 2297 B.C., and more than 4,100 years ago, the Great Yu was appointed 'Superintendent of Public Works' and entrusted with draining off the flood waters and canalizing the river.

Here also was the beginning of the Boxer uprising. Tsingtao sits at the entrance of Kiaochow Bay. Following the war of Japan with China this was seized by Germany, November 14th, 1897, nominally to indemnify for the murder of two German missionaries which had occurred in Shantung, and March 6th, 1898, this bay, to the high water line, its islands and a 'Sphere of Influence' extending 30 miles in all directions from the boundary, together with Tsingtao, was leased to Germany for ninety-nine years. Russia demanded and secured a lease of Port Arthur at the same time. Great Britain obtained a similar lease of Wei-hai-wei in Shantung, while to France Kwangchow-wan in southern China was leased. But the 'encroachments' of European powers did not stop with these leases and during the latter part of 1898 the 'Policy of Spheres of Influence' culminated in the international rivalry for railway concessions and mining. These greatly alarmed China and, as might have been expected of a patriotic, even though naturally peaceful people, they determined to defend their country against such encroachments and the Boxer troubles followed.

Tsingtao has a deep, commodious harbour always free from ice. Here Germany is constructing very extensive and substantial harbour improvements which will be of lasting benefit to the province and the Empire. A pier 4 miles in length encloses the inner wharf, and a second wharf is nearing completion. Germany is also maintaining a meteorological observatory and has established a large, comprehensive Forest Garden, which is showing remarkable developments for so short a time.

Our steamer entered the harbour during the night and, on going ashore, we soon found that only Chinese and German were generally spoken. The afternoon was spent at the Forest Garden and on the reforestation tract, which are under the supervision of Mr. Haas. The Forest Garden covers 270 acres and the reforestation tract 3,000 acres more. In the garden a great variety of forest trees, fruit trees and small fruits are being tried out with high promise of the most valuable results.

It was in the steep hills about Tsingtao that we first saw at close range serious soil erosion in China; and the renewal of forest growth on hills nearly devoid of soil was here remarkable, in view of the long dry seasons which prevail from November to June. Fig. 102 shows how destitute of soil the crests of granite hills may become and yet how the return of forest growth may hasten as soon as it is no longer cut away. The rock going into decay, where this view was taken, is an extremely coarse crystalline granite, as may be seen in contrast with the watch, and it is falling into decay at a marvellous rate. Disintegration has penetrated the rock far below the surface and the large crystals are held together with but little more tenacity than prevails in a bed of gravel. Moisture and even roots penetrate it deeply and readily and the crystals fall apart with thrusts of the knife blade. Roadways have been extensively carved along the sides of the hills with the aid of only pick and shovel. Close examination of the rock shows that layers of sediment exist between the crystal faces, either washed down by percolating rain or formed through decomposition of the crystals in place. The next illustration, Fig. 103, shows how large the growth on such soils may be, and in Fig. 104 the vegetation and forest growth are seen coming back, closely covering just such soil surfaces and rock structure as are indicated in Figs. 102 and 103.

These views are taken on the reforestation tract at Tsingtao, but most of the growth is volunteer, protected now by the German Government in their effort to see what may be possible under careful supervision.

FIG. 102. – Granite hill destitute of soil, rapidly falling into decay. Reforestation area, Tsingtao, Shantung.

The loads of pine bough fuel represented in Fig. 65 were gathered from such hills and from such forest growth as are here represented, but on lands more distant from the city. But Tsingtao, with its 40,000 Chinese, and Kiaochow across the bay, with its 120,000 more, and other villages dotting the narrow plains, maintain a very great demand for such growth

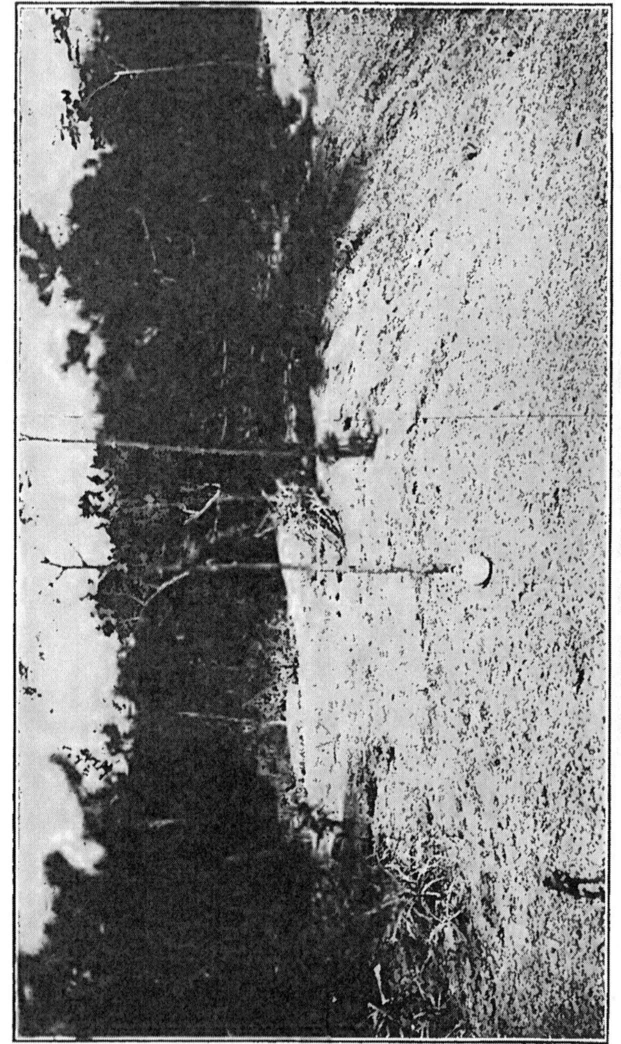

FIG. 103. – Almost soil-free granite surface on which young forest growth, largely pine, is developing. Reforestation tract, Tsingtao, Shantung.

on the hill lands. The wonder is that forest growth has persisted at all and has contributed so much in the way of fuel.

Growing in the Forest Garden was a most beautiful wild yellow rose, native to Shantung. It was used for landscape effect in the parking, and ought to be widely introduced into other countries wherever it will thrive. It was growing as heavy borders and massive clumps 6 to 8 feet high, giving a wonderful effect, with its brilliant, dense cloud of the richest yellow bloom. The blossoms are single, fully as large as the Rosa rugosa. The tips of the petals shade into dainty light straw yellow, while the centre is a deep

FIG. 104. – Forest and herbaceous growth coming back over such soil conditions as are seen in Figs. 102 and 103. Reforestation tract, Tsingtao, Shantung.

orange, with a contrast sufficient to show in the photograph from which Fig. 105 was prepared. Another beautiful and striking feature of this rose is the clustering of the blossoms in one-sided wreath-like sprays, sometimes 12 to 18 inches long, the flowers standing close enough even to overlap.

The next morning we took the early train for Tsinan to obtain a general view of the country and to note the places most favourable as points for field study. We had resolved also to make an effort to secure an interpreter through the Union Medical College at Tsinan. Leaving Tsingtao, the train skirts around the Kiaochow bay for a distance of nearly 50 miles, passing the city of the same name, with its population of 120,000 and an import and export trade valued in 1905 at over $24,000,000. At Sochen we passed through a coal-mining district where coal was being brought to the cars in baskets carried by men. The coal on the loaded open cars was sprinkled with whitewash, which served the purpose of a seal to safeguard against stealing during transit. It was sprinkled in such a way that no coal could be removed without

Fig. 105. – Close view of the wild yellow Shantung rose cultivated in the Forest Garden at Tsingtao, and very effective for parks and pleasure drives.

breaking the pattern of the whitewash and so revealing the theft. This practice is general in China and is applied to many commodities handled in bulk. We saw baskets of milled rice carried by coolies sealed with a pattern laid over the surface by sprinkling some coloured powder upon it. Cut stone, corded for the market, was whitewashed in the same manner as the coal.

As we were approaching Weihsien, another city of 100,000 people, we saw half a dozen teams passing along one of the deeply depressed, centuries-old roadways, worn 8 to 10 feet deep. We had passed several such roads and were puzzled to account for such peculiar erosion. The teams gave the explanation and thus connected our earlier reading with the concrete fact. Along these deep-cut roadways caravans may pass, winding through the fields, entirely unobserved unless one chances to be close along the line or the movement is discovered by clouds of dust.

Weihsien is near one of the great commercial highways of China and in the centre of one of the coal-mining regions of the province. Still further along towards Tsinan we passed Tsingchowfu, another of the large cities of the province, with 50,000 population. All day we rode through fields of wheat, always planted in rows, and in hills in the row east of Kaumi, but in single or double continuous drills westward from here to Tsinan. Thousands of wells used for irrigation, of the type seen in Fig. 107, were passed during the day, many of them recently dug to supply water for the barley then suffering from severe drought.

It was 6.30 p.m. before our train pulled into the station at Tsinan; 7.30 when we had finished supper and engaged a ricksha to take us to the College in quest of an interpreter. We could not speak Chinese, the ricksha boy could neither speak nor understand a word of English, but the hotel proprietor had instructed him where to go. We plunged into the narrow streets of a great Chinese city, the boy running wherever he could, walking where he must on account of the density of the crowds or the roughness of the stone paving. We had turned many corners, crossed bridges and passed through tunnelled archways in sections of the massive city walls, until it was getting dusk and the ricksha man purchased and lighted a lantern. We were to reach the college in thirty minutes but had been out a full hour. A little later the boy drew up to and held conference with a policeman. The curious of the street gathered about, and it dawned upon us that we were lost in the night in the narrow streets of a Chinese city of 300,000 people. To go further would be useless, for the gates of the mission compound would be locked. We could only indicate by motions our desire to return, but these were not understood. On the train a thoughtful, kindly old German had recognized a stranger in a foreign land and volunteered useful information, cutting from his daily paper an advertisement describing a good hotel. This gave the name of the hotel in German, English and in Chinese characters.

We handed this to the policeman, pointing to the name of the hotel, indicating by motions the desire to return, but apparently he was unable to read in either language and seemed to think we were assuming to direct the way to the college. A man and boy in the crowd apparently volunteered to act as escort for us. The throng parted and we left them, turned more corners into more unlighted narrow alleyways, one of which was too difficult to permit us to ride. The escorts, if such they were, finally left us, but the dark alley led on until it terminated at the blank face, probably of some other portion of the massive city wall we had thrice threaded through lighted tunnels. Here the ricksha boy stopped and turned about, but the light from his lantern was too feeble to permit reading the workings of his mind through his face, and our tongues were both utterly useless in this emergency, so we motioned for him to turn back and by some route we reached the hotel at 11 p.m.

We abandoned the effort to visit the college for the purpose of securing an interpreter, and took the early train back to Tsingtao, reaching that place in time to secure the satisfactory service of Mr. Chu Wei Yung, through the kind offices of Mr. Scott. We had been twice over the road between the two cities, obtaining a general idea of the country and of the crops and field operations at this season. The next morning we took an early train to Tsangkau and were ready to walk through the fields and to talk with the last generations of more than forty unbroken centuries of farmers who, with brain and brawn, have successfully and continuously sustained large families on small areas without impoverishing their soil. Fig. 106 is from a photograph taken in one of these fields. We astonished the old farmer by asking the privilege of holding his plough through one round in his little field, a request which he readily granted. Our furrow was not as well turned as his, nor as well as we could have done with a two-handled Oliver or John Deere, but it was better than the old man had expected and won his respect.

This plough had a good steel point, as a separate, blunt, V-shaped piece, and a mouldboard of cast steel with a good twist which turned the soil well. The standard and sole were of wood and at the end of the beam was a block for gauging the depth of furrow. The cost of this plough, to the farmer, was $2.15, gold, and when the day's work is done it is taken home on the shoulders, even though the distance may be a mile or more, and carefully housed. Chinese tradition states that the plough was invented by Shennung, who lived 2837–2697 B.C. and 'taught the art of agriculture and the medical use of herbs.' He is honoured as the 'God of Agriculture and Medicine.'

Through my interpreter we learned that there were twelve in the man's family, maintained on 15 mow of land, or 2·5 acres, together with his team, consisting of a cow and small donkey,

FIG. 106. – A Shantung plough, simple but effective.

besides two pigs. This is at the rate of 192 people, 16 cows, 16 donkeys and 32 pigs on a 40-acre farm; and of a population density equivalent to 3,072 people, 256 cows, 256 donkeys and 512 swine per square mile of cultivated field.

On another small holding we talked with the farmer standing at the well in Fig. 23, where he was irrigating a little piece of barley 30 feet wide and 138 feet long. He owned and was cultivating but $1\frac{2}{3}$ acres of land, and yet there were ten in his family and he kept one donkey and usually one pig. Here is a maintenance capacity at the rate of 240 people, 24 donkeys and 24 pigs on a 40-acre farm; and a population density of 3,840 people, 384 donkeys and 384 pigs per square mile. His usual annual sales in good seasons were equivalent in value to $73, gold.

In both these cases the crops grown were wheat, barley, large and small millet, sweet potatoes and soy beans or peanuts. Much straw braid is manufactured in the province by the women and children in their homes, and the cargo of the steamer on which we returned to Shanghai consisted almost entirely of shelled peanuts in gunny sacks and huge bales of straw braid destined for the manufacture of hats in Europe and America.

Shantung has only a moderate rainfall, little more than 24 inches annually, and this fact has played an important part in determining the agricultural practices of these very old people. In Fig. 107 is a closer view than Fig. 23 of the farmer watering his little field of barley. The well had just been dug, over 8 feet deep, expressly and solely to water this one piece of grain once, after which it would be filled and the ground planted.

The season had been unusually dry, as had been the one before, and the people were fearing famine. Only 2·44 inches of rain had fallen at Tsingtao between the end of the preceding October and our visit, May 21st, and hundreds of such temporary wells had been or were being dug along both sides of the 250 miles of railway, and nearly all to be filled up again when the crop on the ground was irrigated, to release the land for the crop to follow. The homes are in villages a mile or more apart and often the holdings or rentals are scattered, separated by considerable distances, hence easy portability is the key-note in the construction of this irrigating outfit. The bucket is very light, simply a woven basket waterproofed with a paste of bean flour. The windlass turns like a long spool on a single pin and the standard is a tripod with removable legs. Some wells we saw were 16 or 20 feet deep and in these the water was raised by a cow walking straight away at the end of a rope.

The amount and distribution of rainfall in this province, as indicated by the mean of ten years' records at Tsingtao, obtained at the German Meteorological Observatory through the courtesy of Dr. B. Meyermanns, are given in the table in which the rainfall of Madison, Wisconsin, is inserted for comparison.

Fig. 107. – Temporary well and portable irrigation outfit, Shantung.

	Mean monthly rainfall.		Mean rainfall in 10 days.	
	Tsingtao, Inches.	Madison, Inches.	Tsingtao, Inches.	Madison, Inches.
January	·394	1·56	·131	·520
February	·240	1·50	·080	·500
March	·892	2·12	·297	·707
April	1·240	2·52	·413	·840
May	1·636	3·62	·545	1·207
June	2·702	4·10	·901	1·366
July	6·637	3·90	2·212	1·300
August	5·157	3·21	1·719	1·070
September	2·448	3·15	·816	1·050
October	2·258	2·42	·753	·807
November	·396	1·78	·132	·593
December	·682	1·77	·227	·590
Total	24·682	31·65		

While Shantung receives less than 25 inches of rain during the year, against Wisconsin's more than 31 inches, the rainfall during June, July and August in Shantung is nearly 14·5 inches, while Wisconsin receives but 11·2 inches. This greater summer rainfall, with persistent fertilization and intense management, in a warm latitude, are some of the elements permitting Shantung to-day to feed 38,247,900 people from an area equal to that upon which Wisconsin is yet feeding but 2,333,860. Must American agriculture ultimately feed sixteen people where it is now feeding but one? If so, correspondingly intense and effective practices must follow, and we can neither know too well nor too early what these Old World people have been driven to do, how they have succeeded, and how we and they may improve upon their practices and lighten human burdens by more fully utilizing physical forces and mechanical appliances.

As we passed on to other fields we found a mother and daughter transplanting sweet potatoes on carefully fitted ridges of nearly air-dry soil in a little field, the remnant of a table on a deeply eroded hillside (Fig. 108). The husband was bringing water for moistening the soil from a deep ravine a quarter of a mile distant, carrying it on his shoulder in two buckets (Fig. 109) across an intervening gulch. He had excavated four holes at intervals up the gulch and from these, with a broken gourd dipper mended with stitches, he filled his pails, bailing in succession from one to the other in regular rotation.

The daughter was transplanting. Holding the slip with its tip between thumb and fingers, a strong forward stroke ploughed a furrow in the mellow, dry soil; then, with a backward movement and a downward thrust, she planted the slip, firmed the soil about it, leaving a depression in which the mother poured about a pint of water from another gourd dipper. After the water had soaked away, dry earth was drawn about the slip and firm and

FIG. 108. – Strong erosion in Shantung, with wheat on remnants of tables.

looser earth drawn over this, the only tools being the naked hands and dipper.

The father and mother were dressed in coarse garb but the daughter was neatly clad, with delicate hands decorated with rings and a bracelet. Neither of the women had bound feet. There were ten in the family; and on adjacent similar areas they had small patches of wheat nearly ready for the harvest, all planted in hills, hoed, and in astonishingly vigorous condition considering the extreme drought which prevailed. The potatoes were being planted under these extreme conditions in anticipation of the rainy season which then was fully due. The summer before had been one of unusual drought, and famine was threatened. The government had recently issued an edict that no sheep should be sold from the province, fearing they might be needed for food. An old woman in one of the villages came out, as we walked through, and inquired of my interpreter if we had come to make it rain. Such was the stress under which we found these people.

One of the large farmers, owning 10 acres, stated that his usual yield of wheat in good season was 160 catty per mow, equivalent to 21·3 bushels per acre. He was expecting the current season not more than one-half this amount. As a fertilizer he used a

FIG. 109. – Getting water to transplant sweet potatoes. A Standard Oil can is balanced against China's ancient stone jar.

prepared earth compost which we shall describe later, mixing it with the grain and sowing in the hills with the seed, applying about 5,333 pounds per acre, which he valued, in our currency, at $8.60, or $3.22 per ton. A pile of such prepared compost is seen in Fig. 110, ready to be transferred to the field. The views show with what cleanliness the yard is kept and with what care all animal waste is saved. The cow and donkey are the work team, such as was being used by the ploughman referred to in Fig. 106. The mounds in the background of the lower view are graves; the fence behind the animals is made from the stems of the large millet,

FIG. 110. – Two views of the same farmyard, showing a pile of prepared compost and the farm team.

kaoliang, while that at the right of the donkey is made of earth, both indicative of the scarcity of timber. The buildings, too, are thatched and their walls are of earth plastered with an earthen mortar worked up with chaff.

In another field a man ploughing and fertilizing for sweet potatoes had brought to the field and laid down in piles the finely pulverized dry compost. The father was ploughing; his son of sixteen years was following and scattering, from a basket, the pulverized dry compost in the bottom of the furrow. The next furrow covered the fertilizer, four turned together forming a ridge upon which the potatoes were to be planted after a second and older son had smoothed and fitted the crest with a heavy hand rake. The fertilizer was thus applied directly beneath the row, at the rate of 7,400 pounds per acre, valued at $7.15, our currency, or $1.93 per ton.

We were astonished at the moist condition of the soil turned, which was such as to pack in the hand notwithstanding the extreme drought prevailing and the fact that standing water in the ground was more than 8 feet below the surface. The field had been without crop and cultivated.

To the question, 'What yield of sweet potatoes do you expect from this piece of land?' he replied, 'About 4,000 catty,' which is 440 bushels of 56 pounds per acre. The usual market price was stated to be $1.00, Mexican, per 100 catty, making the gross value of the crop $79.49, gold, per acre. His land was valued at $60, Mexican, per mow, or $154.80 per acre, gold.

My interpreter informed me that the average well-to-do farmers in this part of Shantung own from 15 to 20 mow of land, and that this amount is ample to provide for eight people. Such farmers usually keep 2 cows, 2 donkeys, and 8 or 10 pigs. The less well-to-do or small farmers own 2 to 5 mow and act as superintendents for the larger farmers. Taking the largest holding, of 20 mow per family of eight people, as a basis, the density per square mile would be 1,536 people, and an area of farm land equal to the state of Wisconsin would have 86,000,000 people; 21,500,000 cows; 21,500,000 donkeys and 86,000,000 swine. These observations apply to one of the most productive sections of the province, but very large areas of land in the province are not cultivable and the last census showed the total population nearly one-half of this amount. It is clear, therefore, that either very effective agricultural methods are practised or else extreme economy is exercised. Both are true.

On this day in the fields our interpreter procured his dinner at a farm-house, bringing us four boiled eggs, for which he paid at the rate of 8·3 cents of our money, but his dinner was probably included in the price. The next table gives the prices for some articles obtained by inquiry at the Tsingtao market, May 23rd, 1909, reduced to our currency.

	Cents
Old potatoes, per lb.	2·18
New potatoes, per lb.	2·87
Salted turnip, per lb.	·86
Onions, per lb.	4·10
Radishes, bunch of 10	1·29
String beans, per lb.	11·46
Cucumbers, per lb.	5·73
Pears, per lb.	5·73
Apricots, per lb.	8·60
Pork, fresh, per lb.	10·33
Fish, per lb.	5·73
Eggs, per dozen	5·16

The only items which are low compared with our own prices are salted turnips, radishes and eggs. Most of the articles listed were out of season for the locality and were imported for foreigners, turnips, radishes, pork, fish and eggs being the exceptions. Prof. Ross informs us that he found eggs selling in Shensi at four for 1 cent of our money.

Our interpreter asked a compensation of $1, Mexican, or 43 cents, U.S. currency, per day, he furnishing his own meals. The usual wage for farm labour here was $8.60 per year, with board and lodging. We have referred to the wages paid by missionaries for domestic service. As servants the Chinese are considered efficient, faithful and trustworthy. It was the custom of Mr. and Mrs. League to entrust them with the purse for marketing, feeling that they could be depended upon for the closest bargaining. Commonly, when instructed to procure a certain article, if they found the price one or two cash higher than usual they would select a cheaper substitute. If questioned as to why instructions were not followed the reply would be 'Too high, no can afford.'

Mrs. League recited her experience with her cook regarding his use of our kitchen appliances. After fitting the kitchen with a modern range and cooking utensils, and working with him to familiarize him with their use, she was surprised, on going into the kitchen a few days later, to find that the old Chinese stove had been set on the range and the cooking done with the usual Chinese furniture. When asked why he was not using the stove, his reply was 'Take too much fire.' Nothing jars on the nerves of these people more than incurring needless expense, extravagance in any form, or poor judgment in making purchases.

Daily we became more and more impressed by the evidence of the intense and incessant stress imposed by the dense population through centuries, and how, under it, the laws of heredity have wrought upon the people, affecting constitution, habits and character. Even the cattle and sheep have not escaped its irresistible power. Many times in this province we saw men herding flocks of 20 to 30 sheep along the narrow unfenced pathways winding through the fields, and on the grave lands. The prevailing drought had left very little green to be had from these places and yet sheep were literally brushing their sides against fresh green wheat and barley, never molesting them. Time and again the flocks were stampeded into the grain by an approaching train, but immediately they returned to their places without taking a nibble. The voice of the shepherd and an occasional well aimed lump of earth only were required to bring them back to their uninviting pastures.

In Kiangsu and Chekiang provinces a line of half a dozen white goats were often seen feeding single file along the pathways, held by a cord like a string of beads, sometimes led by a child. Here, too, one of the most common sights was the water buffalo grazing unattended among the fields along the paths and canal banks, with crops all about. One of the most memorable shocks came to us in Chekiang, China, when we had fallen into a reverie while gazing at the shifting landscape from the doorway of our low-down Chinese houseboat. Something in the sky and the vegetation along the canal bank had recalled the scenes of boyhood days and it seemed, as we looked aslant up the bank with its fringe of grass, that we were gliding along Whitewater creek through familiar meadows and that standing up would bring the old home in sight. That instant there glided into view, framed in the doorway and projected high against the tinted sky above the setting sun, a giant water buffalo standing motionless as a statue on the summit of a huge grave mound, lifted fully 10 feet above the field. But in a flash this was replaced by a companion scene, with its beautiful setting, which had been as suddenly fixed on the memory fourteen years before in the far-away Trossachs, when our coach, hurriedly rounding a sharp turn in the hills, suddenly exposed a wild ox of Scotland similarly thrust against the sky from a small but isolated rocky summit, and then, outspeeding the wireless, recollection crossed two oceans and an intervening continent, and brought us back to China before a speed of 5 miles per hour could move the first picture across the narrow doorway.

It was through the fields about Tsangkow that the stalwart freighters referred to earlier (Fig. 27) passed us on one of the paths leading from Kiaochow through unnumbered country villages, already 11 miles on their way with their wheelbarrows loaded with matches made in Japan. Many of the wheelbarrow men seen in Shanghai and other cities are from Shantung families. During the harvest season, too, many of these people go west and north into Manchuria seeking employment, returning to their homes in winter.

Alexander Hosie, in his book on Manchuria, states that from Chefoo alone more than 20,000 Chinese labourers cross to Newchwang every spring by steamer, others finding their way there by junks or other means, so that after the harvest season 8,000 more return by steamer to Chefoo than left by that route in the spring, from which he concludes that Shantung annually supplies Manchuria with agricultural labour to the extent of 30,000 men.

The average condition of wheat in Shantung during this dry season, and nearing maturity, is seen in Fig. 111. It stands rather more than 3 feet high, as indicated by our umbrella between the rows. Beyond the wheat and to the right, grave mounds serrate the sky-line. No hills were in sight, for we were in the broad plain built up from the sea between the two mountain islands forming the highlands of Shantung.

On May 22nd we were in the fields north of Kiaochow, some 60 miles by rail west from Tsingtao, but within the neutral zone extending 30 miles back from the high water line of the bay of the same name. Here the Germans had built a broad macadam road after the best European type, but over it were passing the vehicles of forty centuries ago as seen in Fig. 112. It is doubtful if the resistance to travel experienced by these men on the better road is sufficiently less than that on the old paths to convince them that the cost of construction and maintenance is worth while until vehicles and the price of labour change. It may appear strange that with a nation of so many millions and with so long a history, roads have persisted as little more than beaten footpaths; but modern methods of transportation remained physical impossibilities to every people until the science of the last century opened

FIG. 111. – Field of wheat in Shantung, nearing maturity in a season of unusual drought.

the way. Throughout their history the burdens of these people have been carried largely on foot, mostly on the feet of men, and of single men wherever the load could be advantageously divided. Animals have been supplemental burden bearers, but, as with the men, they have carried the load directly on their own feet, the mode least disturbed by inequalities of road surface.

For adaptability to the worst road conditions no vehicle equals the wheelbarrow, progressing by one wheel and two feet. No vehicle is used more in China, if the carrying pole is excepted, and no wheelbarrow in the world permits so high an efficiency of human power as the Chinese, as must be clear from Fig. 27, where nearly the whole load is balanced on the axle of a high, massive wheel with broad tyre. A shoulder band from the handles of the barrow relieves the strain on the hands and, when the load or the road is heavy, men or animals may aid in drawing, or even, when the wind is favourable, it is not unusual to hoist a sail to gain propelling power. It is only in northern China, and then in the more level portions, where there are few or no canals, that carts

FIG. 112. – The vehicles and other means of transport as used for centuries now travelling on a modern road of German construction, Kiaochow. See also Fig. 71.

are extensively used. Most of the heavy carts, especially those in Manchuria, seen in Fig. 182, have the wheels framed rigidly to the axle which revolves with them, the bearing being in the bed of the cart.

In the development and utilization of inland waterways no people have approached the Chinese. In land transportation they have clearly followed the line of least resistance for individual initiative, so characteristic of industrial China.

There are Government courier or postal roads which connect

Peking with the most distant parts of the Empire – twenty-one are usually enumerated. These, as far as practicable, take the shortest route, are often cut into the mountain sides and even pass through tunnels. In the plains regions these roads may be 60 to 75 feet wide, paved and occasionally bordered by rows of trees. In some cases, too, signal towers are erected at intervals of 3 miles, and there are inns along the way, relay posts and stations for soldiers.

We have spoken of planting grain in rows and in hills in the

FIG. 113. – Wheat planted in hills and in rows, the pairs of rows being 30 inches apart and the hills 16 inches, covering 5 feet.

row. In Fig. 113 is a field with the rows planted in pairs, the members being 16 inches apart, and together occupying 30 inches. The space between each pair is also 30 inches, making 5 feet in all. This makes practicable frequent hoeing, begun early in the spring and repeated after every rain. It also makes possible the feeding of the plants when they can utilize food to the best advantage, and the repetition of the feeding if desirable. Besides, the ground in the wider space may be fitted and fertilized, and another crop planted before the first is removed. The hills alternate in the rows, and are 24 to 26 inches from centre to centre.

The planting may be done by hand or with a drill such as that in Fig. 114, ingenious in the simple mechanism which permits planting in hills. The husbandman had just returned from the field with the drill on his shoulder when we met him at the door of his village home. He explained to us the construction and operation of the drill and permitted the photograph to be taken. In the drill there was a heavy-laden weight swinging free from a point above the space between the openings leading to the respective drill feet. When planting, the operator rocks the drill from side to side, causing the weight to hang first over one and then

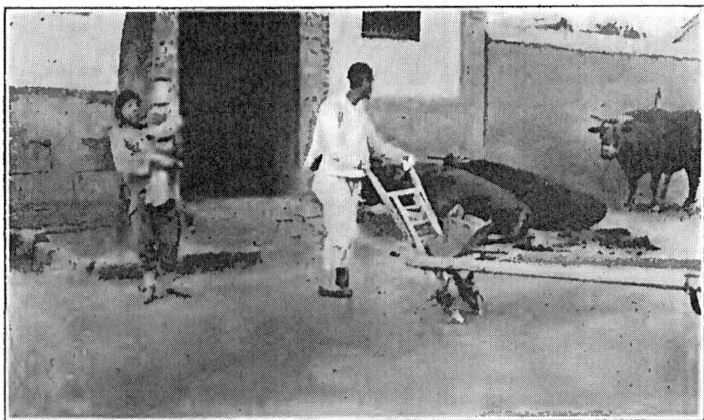

FIG. 114.—Double row seed-drill, just returning from the fields to the village home.

over the other opening, thus securing alternation of hills in each pair of rows.

Counting the heads of wheat in the hill in a number of fields we found that they ranged between 20 and 100, the distance between the rows and between the hills being as stated above. There were always a larger number of stalks per hill where the water capacity of the soil was large, where the ground water was near the surface, and where the soil was evidently of good quality. This may have been partly the result of stooling, but we have little doubt that judgment was exercised in planting, sowing less seed on the lighter soils where less moisture was available. In the piece just referred to, in the illustration, an average hill contained

46 stalks, and the number of kernels in a head varied between 20 and 30. Taking Richardson's estimate of 12,000 kernels of wheat to the pound, this field would yield about 12 bushels of wheat per acre in this unusually dry season. Our interpreter, whose parents lived near Kaomi, four stations further west, stated that in 1901, one of their best seasons, farmers there secured yields as high as 875 catty per legal mow, which is at the rate of 116 bushels per acre. Such a yield on small areas highly fertilized and carefully tilled, when the rainfall is ample or where irrigation is practised, is quite possible, and in the Kiangsu province we observed individual small fields which would certainly approach close to this figure.

Further along in our journey of the day we came upon a field where three, one of them a boy of fourteen years, were hoeing and thinning millet and maize. The usual yield of maize was set at 420 to 480 catty per mow, and that of millet at 600 catty, or 60 to 68·5 bushels of maize and 96 bushels of millet, of 50 pounds, per acre, and the usual price would make the gross earnings $23.48 to $26.83 per acre for the maize, and $30 96, gold, for the millet.

It was evident when walking through these fields that the autumn-sowed grain was standing the drought far better than the barley planted in the spring, quite likely because of the deeper and stronger development of root system made possible by the longer period of growth, and partly because the wheat had made much of its growth earlier and utilized water that had fallen before the barley was planted. Farmers here are very particular to hoe their grain, beginning in the early spring, and always after rains, thoroughly appreciating the efficiency of earth mulches. Their hoe, seen in Fig. 115, is peculiarly well adapted to its purpose, the broad blade being so hung that it draws nearly parallel with the surface, cutting shallow and permitting the soil to drop practically upon the place from which it was loosened. These hoes are made in three parts; a wooden handle, a long, strong and heavy iron socket shank, and a blade of steel. The blade is detachable and different forms and sizes of blades may be used on the same shank. The mulch-producing blades may have a cutting edge 13 inches long and a width of 9 inches.

At short intervals on either hand, along the 250 miles of railway between Tsingtao and Tsinan, were observed many piles of earth compost distributed in the fields. One of these piles is seen in Fig. 116. They were sometimes on unplanted fields, in other cases they occurred among the growing crops soon to be harvested, or where another crop was to be planted between the rows of one already on the ground. Some of these piles were 6 feet high.

Fig. 115. – Method of using the broad, heavy hoe in producing surface mulch, as seen in Shantung.

All were built in cubical form with flat top and carefully plastered with a layer of earth mortar which sometimes cracked on drying, as seen in the illustration. The purpose of this careful shaping and plastering we did not learn, although our interpreter stated it was to prevent the compost from being appropriated for use on adjacent fields. Such a finish would have the effect of a seal, showing if the pile had been disturbed, but we suspect other advantages are sought by the treatment which involves so large an amount of labour.

The amount of this earth compost prepared and used annually in Shantung is large, as indicated by the cases cited, where more than 5,000 pounds, in one instance, and 7,000 pounds in another, were applied per acre for one crop. When two or more crops are grown the same year on the same ground, each is fertilized, hence from 3 to 6 or more tons may be applied to each cultivated acre. The methods of preparing compost and of fertilizing in Kiangsu,

Fig. 116. – Carefully plastered earth compost stacked in the field awaiting distribution, Shantung.

Chekiang and Kwangtung provinces have been described. In this part of Shantung, in Chihli and north in Manchuria as far as Mukden, the methods are materially different and if possible even more laborious, but clearly rational and effective. Here nearly all if not all fertilizer compost is prepared in the villages and carried to the fields, however distant these may be.

Rev. T. J. League very kindly accompanied us to Chengyang on the railway, from which we walked some 2 miles back to a prosperous rural village to see their methods of preparing this compost fertilizer. It was toward the close of the afternoon before we reached the village, and from all directions husbandmen were returning from the fields, some with hoes, some with ploughs, some with drills over their shoulders and others leading donkeys or cattle, and similar customs obtain in Japan, as seen in Fig. 117. These were mostly the younger men. When we reached the village streets the older men, all bare-headed, as were those returning from the fields, and usually with their queues tied about the crown, were chatting, enjoying their pipes of tobacco.

In the matter of conservation of national resources here is one

FIG. 117. – Home after the day's work, in Japan.

of the greatest opportunities open to all civilized nations. What might not be done in the United States with a fund of $57,000,000 annually, the market price of the raw tobacco leaf, and the land, the labour and the capital expended in getting the product to the men who puff, breathe and perspire the noxious product into the air everyone must breathe, and who bespatter the streets, sidewalks, the floor of every public place and conveyance, and befoul the million spittoons, smoking-rooms and smoking-cars, all unnecessary and should be uncalled for, but whose installation and upkeep the non-user as well as the user is forced to pay for, and this in a country of, for and by the people. This costly, filthy, selfish tobacco habit should be outgrown. Let it begin in every new home, where the mother helps the father in refusing to set the example, and let its indulgence be absolutely prohibited to every one while in public school and to all in educational institutions.

Mr. League had been given a letter of introduction to one of the leading farmers of the village, and it chanced that as we reached the entranceway to his home we were met by his son, just returning from the fields with his drill on his shoulder, and it is he standing in the illustration, Fig. 114, holding the letter of introduction in his hand. After we had taken this photograph and another one looking down the narrow street from the same point, we were led to the small open court of the home, perhaps 40 by 80 feet, upon which all doors of the one-storied structures opened. It was dry and bare of everything green, but a row of very tall handsome trees, close relatives of our cottonwood, with trunks 30 feet to the limbs, looked down into the court over the roofs of the low thatched houses. Here we met the father and grandfather of the man with the drill, so that, with the boy carrying the baby in his arms, who had met his father in the street gateway, there were four generations of males at our conference. There were women and girls in the household, but custom requires them to remain in retirement on such occasions.

A low narrow four-legged bench, not unlike our carpenter's saw-horse, 5 feet long, was brought into the court as a seat, which our host and we occupied in common. We had been similarly received at the home of Mrs. Wu in Chekiang province. On our right was the open doorway to the kitchen in which stood, erect and straight, the tall spare figure of the patriarch of the household, his eyes still shining black but with hair and long thin straggling beard a uniform dull ashen grey. He seemed to have assumed the duties of cook, for while we were there he lighted the fire in the kitchen and was busy, but was always the final oracle on any matter of difference of opinion between the younger men regarding answers to questions. Two sleeping apartments adjoining the kitchen, through whose wide kang beds the waste heat from the cooking was conveyed, as described on page 127, completed this side of the court. On our left was the main street completely shut off by a solid earth wall as high as the eaves of the house, while in front of us, adjoining the street, was the manure midden, a compost pit 6 feet deep and some 8 feet square. A low opening in the street wall permitted the pit to be emptied and to receive earth and stubble or refuse from the fields for composting. Against the pit and without partition, but cut off from the court, was the home of the pigs, both under a common roof continuous with a closed structure joining with the sleeping apartments, while behind us and along the alleyway by which we had entered were other dwelling and storage compartments. Thus was the large family of four generations provided with a peculiarly private open court where they could work and come out for sun and air, both, from our standards, too meagrely provided in the houses.

We had come to learn more of the methods of fertilizing practised by these people. The manure midden was before us and the piles of earth brought in from the fields, for use in the process, were stacked in the street, where we had photographed them at the entrance, as seen in Fig. 118. There a father, with his pipe, and two boys stand at the extreme left; beyond them is a large pile of earth brought into the village and carefully stacked in the narrow street; on the other side of the street, at the corner of the first building, is a pile of partly fermented compost thrown from a pit behind the walls. Further along in the street, on the same side, is a second large stack of soil where two boys are standing at either end. In front of the tree, on the left side of the street, stands a third boy, near him a small donkey and still another boy. Beyond this boy stands a third large stack of soil, while still beyond and across the way is another pile partly composted. Notwithstanding the cattle in the preceding illustration, the donkey, the men, the boys, the three long high stacks of soil and the two piles of compost, the ten rods of narrow street possessed a width of available travel-way and a cleanliness which would appear impossible. Each farmer's household had its stack of soil in the street, and in walking through the village we passed dozens of men turning and mixing the soil and compost, preparing it for the field.

The compost pit in front of where we sat was two-thirds filled. In it had been placed all the manure and waste of the household and street, all stubble and waste roughage from the field, all

FIG. 118. – Farm village street with stacks of earth and piles of compost for use as fertilizer, Shantung.

ashes not to be applied directly, and some of the soil stacked in the street. Sufficient water was added at intervals to keep the contents completely saturated and nearly submerged, the object being to control the character of fermentation taking place.

The capacity of these compost pits is determined by the amount of land served, and the period of composting is made as long as possible, the aim being to have the fibre of all organic material completely broken down, the result being a product of the consistency of mortar.

When it is near the time for applying the compost to the field, or of feeding it to the crop, the fermented product is removed in waterproof carrying baskets to the floor of the court, to the yard, such as seen in Fig. 110, or to the street, where it is spread to dry, to be mixed with fresh soil, more ashes, and repeatedly turned and stirred to bring about complete aeration and to hasten the processes of nitrification. During all these treatments, whether in the compost pit or on the nitrification floor, the fermenting organic matter in contact with the soil is converting plant food elements into soluble plant food substances in the form of potassium, calcium and magnesium nitrates and soluble phosphates of one or another form, perhaps of the same bases and possibly others of organic type. If there is time and favourable temperature and moisture conditions for these fermentations to take place in the soil of the field before the crop will need it, the compost may be carried direct from the pit to the field and spread broadcast, to be ploughed under. Otherwise the material is worked and reworked, with more water added if necessary, until it becomes a rich complete fertilizer, allowed to become dry and then finely pulverized, sometimes using stone rollers drawn over it by cattle, the donkey or by hand. The large numbers of stacks of compost seen in the fields between Tsingtao and Tsinan were of this type. They had been laboriously prepared in the villages before being transported to the fields, stacked and plastered, to be ready for use at next planting.

In the early days of European history, before modern chemistry had provided the cheaper and more expeditious method of producing potassium nitrate for the manufacture of gunpowder and fireworks, the nitre-farming to which much land and effort were devoted was no other than a specific application of this most ancient Chinese practice and probably imported from China. While it was not until 1877 to 1879 that men of science came to know that the processes of nitrification, so indispensable to agriculture, are due to germ life, in simple justice to the plain farmers of the world, to those who through all the ages from Adam down, living close to Nature and working through her and with her, have fed the world, it should be recognized that there have been those among them who have grasped these essential vital truths and have kept them alive in the practices of their day. And so we find it recorded in history as far back as 1686 that Judge Samuel Lewell copied upon the cover of his journal a practical man's recipe for making saltpetre beds, in which it was directed, among other things, that there should be added to it 'mother of petre.' It meant, in Judge Lewell's understanding, simply soil from an old nitre bed, but in the mind of the man who applied the maternity prefix, – mother, – it must have meant a vital germ contained in the soil, carried with it, capable of reproducing its kind and of perpetuating its characteristic work, belonging to the same category with the old, familiar, homely germ, 'mother' of vinegar. So, too, with the old cheesemaker who grasped the conception which led to the long-time practice of washing the walls of a new cheese factory with water from an old factory of the same type, he must have been led by analogies of experience with things seen to realize that he was here dealing with a vital factor. Hundreds, of course, have practised empirically, but some one preceded with the essential thought and it is small credit to men of our time who, after ten or twenty years of technical training, having their attention directed to a something to be seen, and armed with compound microscopes which permit them to see with the physical eye the 'mother of petre,' arrogate to themselves the discovery of a great truth. Much more modest would it be and much more in the spirit of giving credit where credit is due to admit that, after long doubting the existence of such an entity, we have succeeded in confirming the truth of a great discovery which belongs to an unnamed genius of the past, or perhaps to a hundred of them who, working with life's processes and familiar with them through long intimate association, saw in these invisible processes analogies that revealed to them the essential truth in such fullness as to enable them to build upon it an unfailing practice.

There is another practice followed by the Chinese, connected with the formation of nitrates in soils, which again emphasizes the national trait of saving and turning to use any and every thing worth while. Our attention was called to this practice by the Rev. A. E. Evans of Shunking, Szechwan province. It rests upon the tendency of the earth floors of dwellings to become heavily charged with calcium nitrate through the natural processes of nitrification. Calcium nitrate being deliquescent absorbs moisture sufficiently to dissolve and make the floor wet and sticky. Dr. Evans' attention was drawn to the wet floor in his own house, which he at first ascribed to insufficient ventilation, but which was unaffected by any improvement in that respect. The father of one of his assistants, whose business consisted in purchasing the soil of such floors for producing potassium nitrate, used so much in China in the manufacture of fireworks and gunpowder, explained his difficulty and suggested the remedy.

This man goes from house to house through the village, purchasing the soil of floors which have thus become overcharged. He procures a sample, tests it and announces what he will pay for the surface 2, 3 or 4 inches, the price sometimes being as high as 50 cents for the privilege of removing the top layer of the floor, which the proprietors must replace. He leaches the soil removed, to recover the calcium nitrate, and then pours the leachings through plant ashes containing potassium carbonate, for the purpose of transforming the calcium nitrate into the potassium nitrate or saltpetre. Dr. Evans learned that during the four months preceding our interview this man had produced sufficient potassium nitrate to bring his sales up to $80, Mexican. It was necessary for him to make a two-days' journey to market his product. In addition he paid a licence fee of 80 cents per month. He must purchase his fuel ashes and hire the services of two men.

When the nitrates which accumulate in the floors of dwellings are not collected for this purpose the soil goes to the fields to be used directly as a fertilizer, or it may be worked into compost. In the course of time the earth used in the village walls and even in the construction of the houses may disintegrate so as to require removal, but in all such cases, as with the earth brick used in the kangs, the value of the soil has improved for composting and is generally so used. This improvement of the soil will not appear strange when it is stated that such materials are usually from the subsoil, whose physical condition would improve when exposed to the weather, converting it in fact into an uncropped virgin soil.

We were unable to secure definite data as to the chemical composition of these composts and cannot say what amounts of available plant food the Shantung farmers are annually returning to their fields. There can be little doubt, however, that the amounts are quite equal to those removed by the crops. The soils appeared well supplied with organic matter and the colour of the foliage and the general aspect of crops indicated good feeding.

The family with whom we talked in the village place their usual yields of wheat at 420 catty of grain and 1,000 catty of straw per mow,[1] the grain being worth 35 strings of cash and the straw 12 to 14 strings, a string of cash being 40 cents, Mexican, at this time. Their yields of beans were such as to give them a return of 30 strings of cash for the grain and 8 to 10 strings for the straw. Small millet usually yielded 450 catty of grain, worth 25 strings of cash, per mow, and 800 catty of straw worth 10 to 11 strings of cash; while the yields of large millet they placed at 400 catty per mow, worth 25 strings of cash, and 1,000 catty of straw worth 12 to 14 strings of cash. Stating these amounts in bushels per acre and in our currency, the yield of wheat was 42 bushels of grain and 6,000 pounds of straw per acre, having a cash value of $27.09 for the grain and $10.06 for the straw. The soy bean crop follows the wheat, giving an additional return of $23.22 for the beans and $6.97 for the straw, making the gross earning for the two crops $67.34 per acre. The yield of small millet was 54 bushels of seed and 4,800 pounds of straw per acre, worth $27.09 and $8.12 for seed and straw respectively, while the kaoliang or large millet gave a yield of 48 bushels of grain and 6,000 pounds of stalks per acre, worth $19.35 for the grain and $10.06 for the straw.

A crop of wheat like the one stated, if no part of the plant food contained in the grain or straw were returned to the field, would deplete the soil to the extent of about 90 pounds of nitrogen, 15 pounds of phosphorus and 65 pounds of potassium; and the crop of soy beans, if it also were entirely removed, would reduce these three plant food elements in the soil to the extent of about 240 pounds of nitrogen, 33 pounds of phosphorus and 102 pounds

[1] Their mow was four-thirds of the legal standard mow.

of potassium, on the basis of 45 bushels of beans and 5,400 pounds of stems and leaves per acre, assuming that the beans added no nitrogen to the soil, which is of course not true. This household of farmers, therefore, in order to maintain this producing power in their soil, have been compelled to return to it annually, in one form or another, not less than 48 pounds of phosphorus and 167 pounds of potassium per acre. The 330 pounds of nitrogen they would have to return in the form of organic matter or accumulate it from the atmosphere, through the instrumentality of their soy bean crop or some other legume. It has already been stated that they do add more than 5,000 to 7,000 pounds of dry compost, which, repeated for a second crop, would make an annual application of 5 to 7 tons of dry compost per acre annually. They do use, in addition to this compost, large amounts of bean and peanut cake, which carry all the plant food elements derived from the soil which are contained in the beans and the peanuts. If the vines are fed, or if the stems of the beans are burned for fuel, most of the plant food elements will be returned to the field, and they have doubtless learned how to completely restore these elements.

The roads made by the Germans in the vicinity of Tsingtao enabled us to travel by ricksha into the adjoining country, and on one such trip we visited a village mill for grinding soy beans and peanuts in the manufacture of oil, and Fig. 119 shows the stone roller, 4 feet in diameter and 2 feet thick, which is revolved about a vertical axis on a circular stone plate, drawn by a donkey, crushing the kernels partly by its weight and partly by a twisting motion, for the arm upon which the roller revolves is very short. After the meal had been ground the oil was expressed in essentially the same way as that described for the cotton seed, but the bean and peanut cakes are made much larger than the cotton-seed cakes, about 18 inches in diameter and 3 to 4 inches thick. Two of these cakes are seen in Fig. 120, standing on edge outside the mill in an orderly clean court. It is in this form that bean cake is exported in large quantities to different parts of China and Japan, for use as a fertilizer, and very recently to Europe for both stock food and fertilizer.

Nowhere in this province, nor further north, did we see the large terra-cotta receptacles so extensively used in the south for storing human excreta. In these drier climates some method of

Fig. 119. – Stone mill for grinding soy beans and peanuts, Shantung.

desiccation is practised. We found the gardeners in the vicinity of Tsingtao with quantities of the fertilizer stacked under matting shelters in the desiccated condition. The next illustration, Fig. 121, shows one of these piles being fitted for the garden, its

Fig. 120. – Two large peanut cakes and a paper demijohn for containing the oil outside a village mill, Shantung.

thatched shelter standing behind the grandfather of a household. His grandson was carrying the prepared fertilizer to the garden area seen in Fig. 122, where the father was working it into the soil. The greatest pains are taken, both in reducing the product to a fine powder and in spreading and incorporating it with the soil, for one of the maxims of soil management is to make each square foot of field or garden the equal of every other in its power to produce. In this manner each little holding is made to yield the highest returns possible under the conditions the husbandman is able to control.

FIG. 121. – Pulverizing desiccated human excreta preparatory for use in garden fertilization, Shantung.

From one portion of the area being fitted, a crop of artemisia had been harvested, giving a gross return at the rate of $73.19 per acre, and from another leeks had been taken, bringing a gross return of $13.86 per acre. Chinese celery was the crop for which the ground was now being fitted.

The application of soil as a fertilizer to the fields of China, whether derived from the subsoil or from the silts and organic matter of canals and rivers, must have played an important part in the permanency of agriculture in the Far East, for all such additions have been positive accretions to the effective soil, increasing its depth and carrying to it all plant food elements. If not more than one-half of the weight of compost applied to the fields of Shantung is highly fertilized soil, the rates of application observed would, in a thousand years, add more than 2,000,000 pounds per acre. This represents about the volume of soil we turn with the plough in our ordinary tillage operations, and may carry more than 6,000 pounds of nitrogen, 2,000 pounds of phosphorus and more than 60,000 pounds of potassium.

FIG. 122. – Gardener thoroughly incorporating fertilizer with his soil preparatory to planting a second crop of the season, Shantung.

When we left our hotel by ricksha for the steamer, to return to Shanghai, we observed a boy of thirteen or fourteen years apparently following, sometimes a little ahead, sometimes behind, usually keeping the sidewalk but slackening his pace whenever the ricksha man came to a walk. It was a full mile to the wharf. The boy evidently knew the sailing schedule and judged by the valise in front that we were to take the out-going steamer and that he might possibly earn 2 cents, Mexican, the usual fee for taking a valise aboard the steamer. Twenty men at the wharf might be waiting for the job, but he was taking the chance with the mile down and back thrown in, and all for less than 1 cent in U.S. currency, equivalent at the time to about twenty 'cash.' As we neared the steamer the lad closed up behind, but strong and eager men were watching. Twice he was roughly thrust aside and before the ricksha stopped a man of stalwart frame seized the valise and, had we not observed the boy thus unobtrusively entering the competition, he would have had only his trouble for his pains. Thus intense was the struggle for existence and thus did a mere lad put himself effectively into it. True to breeding and example he had spared no labour to win and was surprised but grateful to receive more than he had expected.

XI

ORIENTALS CROWD BOTH TIME AND SPACE

TIME is a function of every life process, as it is of every physical, chemical and mental reaction, and the husbandman is compelled to shape his operations so as to conform with the time requirements of his crops. The oriental farmer is a time economizer beyond any other. He utilizes the first and last minute and all that are between. The foreigner accuses the Chinaman of being always 'long on time,' never in a fret, never in a hurry. And why should he be when he leads time by the forelock, and uses all there is?

The customs and practices of these Farthest East people in their manufacture of fertilizers in the form of earth composts for their fields, and in their use of altered subsoils which have served in their kangs, village walls and dwellings, are all instances in which they profoundly shorten the time required in the field to affect the necessary chemical, physical and biological reactions. Not only do they thus increase their time assets, but they add, in effect, to their land area by producing these changes outside their fields, at the same time giving their crops the immediately active soil products.

Their compost practices have been of the greatest consequence to them, both in their extremely wet, rice-culture methods, and in their 'dry-farming' methods in parts of the country where the soil moisture is too scanty to permit rapid fermentation under field conditions. Western agriculturists have not sufficiently appreciated the fact that the most rapid growth of plant food substances in the soil cannot occur at the same time and place with the most rapid crop increase, because both processes draw upon the available soil moisture, soil air and soluble potassium, calcium, phosphorus and nitrogen compounds. Whether this fundamental principle of practical agriculture is written in their literature or not it is most indelibly fixed in their practice. If we and they can perpetuate the essentials of this practice at a large saving of human effort, or perpetually secure the final result in some more expeditious and less laborious way, most important progress will have been made.

When we went north to the Shantung province the Kiangsu

and Chekiang farmers were engaged in another of their time-saving practices. This was the planting of cotton in wheat fields before the wheat was quite ready to harvest. In the sections of these two provinces which we visited most of the wheat and barley were sowed broadcast on narrow raised lands, some 5 feet wide, with furrows between, after the manner seen in Fig. 123, in the immediate foreground of which is shown a reservoir on whose bank is installed one of the four-man foot-power irrigation pumps in use to flood the nursery rice bed close by on the right. The narrow lands of broadcasted wheat extend back from the reservoir

Fig. 123. – Looking across reservoir and four-man foot-power pump at fields of grain sowed broadcast in narrow beds.

toward the farmsteads which dot the landscape, and on the left stands one of the pump shelters near the canal bank.

To save time, or lengthen the growing season of the cotton which was to follow, this seed was sown broadcast among the grain on the surface, some ten to fifteen days before the wheat would be harvested. To cover the seed the soil in the furrows between the beds had been spaded loose to a depth of 4 or 5 inches, finely pulverized, and then with a spade evenly scattered over the bed, in such a way as to allow it to sift down among the grain, covering the seed. This loose earth, so applied, acts as a mulch to conserve the capillary moisture, and permit the soil to become sufficiently damp to germinate the seed before the wheat is harvested. The next illustration, Fig. 124, shows our interpreter standing in another field of wheat in which cotton was being sowed in the manner described, although the stand of grain was very close and shoulder high, making it not an easy task either to sow the seed or to scatter sufficient soil to cover it.

When we had returned from Shantung this piece of grain had

Fig. 124. – Field of wheat with grain 4 feet 8 inches high, nearing time of harvest, in which cotton is planted.

been harvested, giving a yield of 95·6 bushels of wheat and 3·5 tons of straw per acre, computed from the statement of the owner that 400 catty of grain and 500 catty of straw had been taken from the beds measuring 4,050 square feet. On the morning of May 29th the photograph for Fig. 125 was taken, showing the same area after the wheat had been harvested and the cotton was up, the young plants showing slightly through the short stubble. These beds had already been once treated with liquid fertilizer. A little later the plants would be hoed and thinned to a stand of about one plant per square foot of surface. There were thirty-seven days between the taking of the two photographs, and certainly thirty days had been added to the cotton crop by this method of planting. over what would have been available if the grain had been first harvested and the field fitted before planting. It will be observed that the cotton follows the wheat without ploughing, but the soil was deep, naturally open, and a layer of nearly two inches of loose earth had been placed over the seed at

FIG. 125. – View of same field as Fig. 124, after the grain had been cut and removed and the cotton sown in it was up.

the time of planting. Besides, the ground would be deeply worked with the two- or four-tined hoe, at the time of thinning.

Starting cotton in the wheat in the manner described is but a special case of a general practice widely in vogue. The growing of multiple crops is the rule throughout these countries wherever the climate permits. Sometimes as many as three crops occupy the same field in recurrent rows, but of different dates of planting and in different stages of maturity. Reference has been made to the overlapping and alternation of cucumbers with greens. The general practice of planting nearly all crops in rows lends itself readily to systems of multiple cropping, and these to the fullest possible utilization of every minute of the growing season and of the time of the family in caring for the crops. In the field, Fig. 126, a crop of winter wheat was nearing maturity, a crop of windsor beans was about two-thirds grown, and cotton had just been planted, April 22nd. This field had been thrown into ridges some 5 feet wide with a 12-inch furrow between them. Two rows of

Fig. 126. – Multiple crops, wheat, windsor beans and cotton. Wheat ready to harvest, beans two-thirds grown, cotton just planted. Upper view looking between wheat rows; lower, looking between bean rows now covering ground.

wheat 8 inches wide, planted 2 feet between centres, occupied the crest of the ridge, leaving a strip 16 inches wide, seen in the upper section, (1) for tillage, (2) then fertilization and (3) finally the row of cotton planted just before the wheat was harvested. Against the furrow on each side was a row of windsor beans, seen in the lower view, hiding the furrow, which was matured some time after the wheat was harvested and before the cotton was very large. A late autumn crop sometimes follows the windsor beans after a period of tillage and fertilization, making four in one year. With such a succession fertilization for each crop and an abundance of soil moisture are required to give the largest returns from the soil.

In another plan winter wheat or barley may grow side by side with a green crop, such as the 'Chinese clover' (*Medicago denticulata*, Willd.) for soil fertilizer, as was the case in Fig. 127, to be turned under and fertilize for a crop of cotton planted in rows on either side of a crop of barley. After the barley had been harvested the

FIG. 127. – Turning under a crop of 'Chinese clover' for green manure, grown with barley and to be followed by cotton.

ground it occupied would be tilled and further fertilized, and when the cotton was nearing maturity a crop of rape might be grown, from which 'salted cabbage' would be prepared for winter use.

Multiple crops are grown as far north as Tientsin and Peking, generally wheat, maize, large and small millet and soy beans, and this, too, where the soil is less fertile and where the annual rainfall is only about 25 inches. Fig. 128 shows one of these fields as it appeared June 14th, where two rows of wheat and two of large millet were planted in alternating pairs, about 28 inches apart. The wheat was ready to harvest but the straw was unusually short because growing on a light sandy loam in a season of exceptional drought.

The piles of pulverized dry-earth compost seen between the

rows had been brought for use on the ground occupied by the wheat when that was removed. The wheat would be pulled, tied in bundles, taken to the village and the roots cut off, for making compost, as in Fig. 129, which shows the family engaged in cutting the roots from the small bundles of wheat, using a long straight knife blade, fixed at one end, and thrust downward upon the bundle with lever pressure. These roots, if not used as fuel, would be transferred to the compost pit in the enclosure seen in Fig. 130, the walls of which were built of earth brick. Here, with any other waste litter, manure or ashes, they would be permitted

Fig. 128. – Multiple crops in Chihli – wheat and sorghum, the wheat ripe, to be followed by soy beans. Piles of compost earth for soy beans.

to decay under water until the fibre had been destroyed, and so reduced to a condition in which it could be incorporated with the soil and applied to the fields. Thus, rich in soluble plant food with nothing to hinder the capillary movement of soil moisture the work proceeded outside the field and the changes could occur unimpeded and without interfering with the growth of crops on the ground.

In this system of combined intertillage and multiple cropping the oriental farmer takes advantage of whatever good may result from rotation or succession of crops, whether these be physical, vito-chemical or biological. If plants are mutually helpful through close association of their root systems in the soil, as some believe may

FIG. 129. – Family engaged in cutting, from bundles of wheat, the roots to be used in making compost, Chihli.

be the case, this growth of different species in close juxtaposition would seem to provide the opportunity, but the other advantages which have been pointed out are so evident and so important that they, rather than this, have doubtless led to the practice described.

FIG. 130. – Compost shelter and pig pen, with pile of wheat roots stacked at one end, for use in making compost, Chihli.

XII

RICE CULTURE IN THE ORIENT

THE basal food crop of the people of China, Korea and Japan is rice, and the mean consumption in Japan, for the five years ending 1906, *per capita* per annum, was 302 pounds. Of Japan's 175,428 square miles she devoted, in 1906, 12,856 to the rice crop. Her average yield of water rice on 12,534 square miles exceeded 33 bushels per acre, and the dry land rice averaged 18 bushels per acre on 321 square miles. In the Hokkaido, as far north as northern Illinois, Japan harvested 1,780,000 bushels of water rice from 53,000 acres.

In Szechwan province, China, Consul-General Hosie places the yield of water rice on the plains land at 44 bushels per acre, and that of the dry land rice at 22 bushels. Data given us in China show an average yield of 12 bushels of water rice per acre, while the average yield of wheat was 25 bushels per acre, the normal yield in Japan being about 17 bushels.

If the rice eaten *per capita* in China proper and Korea is equal to that in Japan the annual consumption for the three nations, using the round number 300 pounds *per capita* per annum, would be:

	Population.	Consumption.
China	410,000,000	61,500,000 tons
Korea	12,000,000	1,800,000 tons
Japan	53,000,000	7,950,000 tons
Total	475,000,000	71,250,000 tons

If the ratio of irrigated to dry land rice in Korea and China proper is the same as that in Japan, and if the mean yield of rice per acre in these countries were 40 bushels for the water rice and 20 bushels for the dry land rice, the acreage required to give this production would be:

	Area. Water rice, sq. miles.	Dry land rice, sq. miles.		
In China	78,073	4,004		
In Korea	2,285	117		
In Japan	12,534	321		
	92,892	4,442	Total	97,334

Our observations along the 400 miles of railway in Korea between Antung, Seoul and Fusan, suggest that the land under rice in this country must be more rather than less than that computed, and the square miles of canalized land in China, as indicated on pages 93 to 98, would indicate an acreage of rice for her quite as large as estimated.

In the three main islands of Japan more than 50 per cent of the cultivated land produces a crop of water rice each year and 7·96 per cent of the entire land area of the Empire, omitting far-north Karafuto. In Formosa and in southern China large areas produce two crops each year. At the large mean yield used in the computation the estimated acreage of rice in China proper amounts to 5·93 per cent of her total area, and this is 7,433 square miles greater than the acreage of wheat in the United States in 1907. Our yield of wheat, however, was but 19,000,000 tons, while China's output of rice was certainly double and probably three times this amount from nearly the same acreage of land. Notwithstanding this large production per acre, more than 50 per cent, possibly as high as 75 per cent, of the same land matures at least one other crop the same year, and much of this may be wheat or barley, both chiefly consumed as human food.

Had the Mongolian races spread to and developed in North America instead of, or as well as, in eastern Asia, there might have been a Grand Canal, somewhat as suggested in Fig. 131, from the Rio Grande to the mouth of the Ohio River and from the Mississippi to Chesapeake Bay. There would thus have been more than 2,000 miles of inland waterway, serving commerce, holding up and redistributing both the run-off water and the wasting fertility of soil erosion, spreading them over 200,000 square miles of thoroughly canalized coastal plains, so many of which are now impoverished lands, through the intolerable waste of a vaunted civilization. And who shall venture to enumerate the increase in the tonnage of sugar, bales of cotton, sacks of rice, boxes of oranges, baskets of peaches, trainloads of cabbage, tomatoes and celery such husbanding would make possible through all time; or number the increased millions they could feed and clothe? We may prohibit the exportation of our phosphorus, grind our limestone, and apply them to our fields, but this alone is only temporizing with the future. The more we produce, the more numerous our millions, the faster must present practices speed the waste to the sea, from whence neither money nor prayer can call them back.

If the United States is to endure; if we are to project our history even through four or five thousand years as the Mongolian nations have done, and if that history is to be written in continuous peace, free from periods of widespread famine or pestilence, this nation must re-orient itself; it must square its practices with a conservation of resources which can make endurance possible.

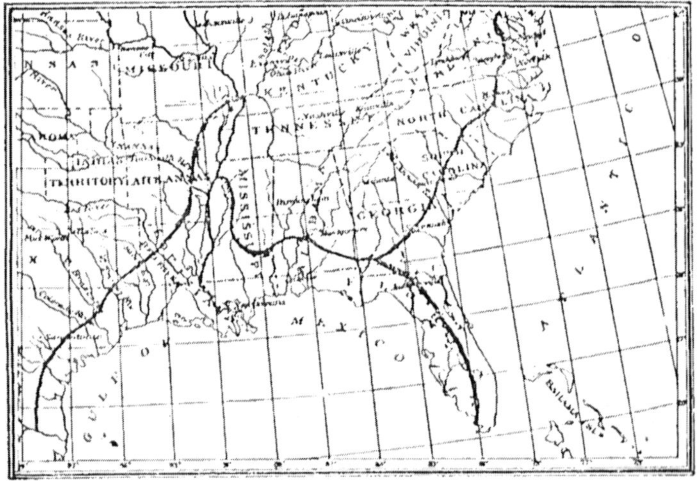

Fig. 131. – A canal which would correspond with the Grand Canal of China.

Intensifying cultural methods but intensifies the digestion, assimilation and exhaustion of the surface soil, from which life springs. Multiple cropping, closer stands on the ground and stronger growth, all mean the transpiration of much more water per acre through the crops, and this can only be rendered possible through a redistribution of the run-off and the adoption of irrigation practices in humid climates where water exists in abundance. Sooner or later we must adopt a national policy which will more completely conserve our water resources, utilizing them not only for power and transportation, but primarily for the maintenance of soil fertility and greater crop production through supplemental irrigation, and all these great national interests must be considered collectively, broadly, and with a view to the fullest possible co-ordination. China, Korea and Japan long ago struck the keynote of permanent agriculture, but the time has now come when they can and will make great improvements, and it remains for us and other nations to profit by their experience, to adopt and adapt what is good in their practice and help in a world movement for the introduction of new and improved methods.

In selecting rice as their staple crop; in developing and maintaining their systems of combined irrigation and drainage, notwithstanding they have a large summer rainfall; in their systems of multiple cropping; in their extensive and persistent use of legumes; in their rotations for green manure to maintain the humus of their soils and for composting; and in the almost religious fidelity with which they have returned to their fields every form of waste which can replace plant food removed by the crops, these nations have demonstrated a grasp of essentials and of fundamental principles which may well cause western nations to pause and reflect.

While this country need not and could not now adopt their laborious methods of rice culture, and while, let us hope, those who come after us may never be compelled to do so, it is nevertheless worth while to study them, for the sake of the principles involved.

Great as is the acreage of land in rice in these countries, but little, relatively, is of the dry land type, and the fields upon which most of the rice grows have all been graded to a water level and surrounded by low, narrow raised rims, such as may be seen in Fig. 132 and in Fig. 133. If the country is not level then the slopes are graded into horizontal terraces varying in size according to the steepness of the areas in which they are cut. We saw these often no larger than the floor of a small room, and Professor Ross informed me that he walked past some in the interior of China no larger than a dining table and one bearing its crop of rice, surrounded by its rim and holding water, yet barely larger than a good napkin. The average area of the paddy field in Japan is officially reported at 1·14 se, or an area of but 31 by 40 feet. Excluding Hokkaido, Formosa and Karafuto, 53 per cent of the

Fig. 132. – Recently transplanted rice fields in Japan.

irrigated rice lands in Japan are in allotments smaller than one-eighth of an acre, and 74 per cent of other cultivated lands are held in areas less than one-fourth of an acre, and each of these may be further subdivided. The next two illustrations, Figs. 134 and 135, give a good idea both of the small size of the rice fields and of the terracing which has been done to secure the water level basins. The house standing near the centre of Fig. 134 makes a good scale for judging both the size of the fields and the slope of

Fig. 133. – Rice fields on the plains of the Yangtse-kiang being flooded preparatory to transplanting rice.

the valley. The distance between the rows of rice is scarcely one foot, so that counting those in the foreground will furnish another means of measuring. There are more than twenty little fields shown in front of the house and reaching but half-way to it, and the house was less than 500 feet from the camera.

There are more than 11,000 square miles of fields thus graded in the three main islands of Japan, each provided with rims, with water supply and drainage channels, all carefully kept in the best of repair. The more level areas, too, in each of the three countries,

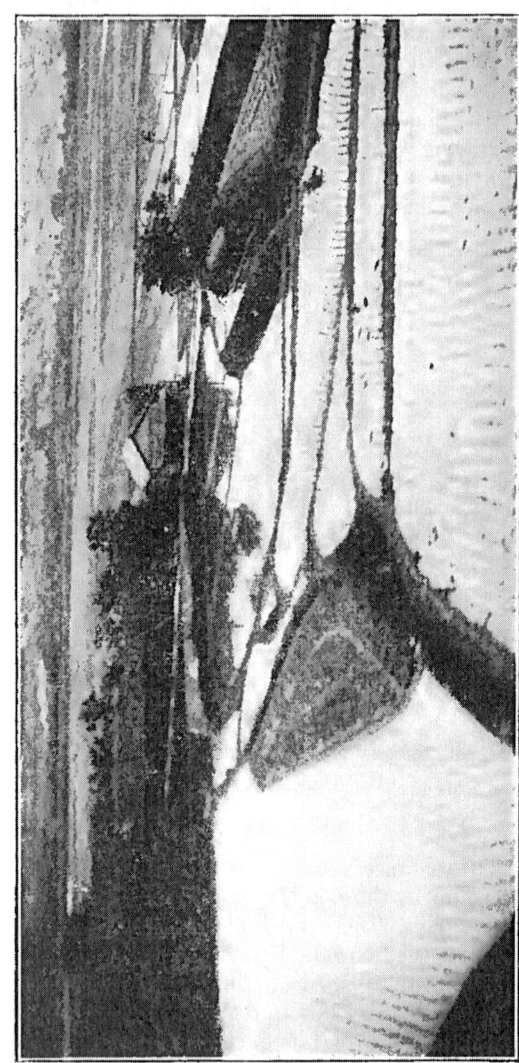

Fig. 134. – Terraced valley with small terraces flooded and transplanted to rice, Japan.

have been similarly thrown into water level basins, comparatively few of which cover large areas, because nearly always the holdings are small. All the earth excavated from the canals and drainage channels has been levelled over the fields unless needed for levees or dikes, so that the original labour of construction, added to that of maintenance, makes a total far beyond our comprehension and nearly all of it is the product of human effort.

FIG. 135. – Looking down a steep, narrow Japanese valley at small, flooded and transplanted rice paddies.

The laying out and shaping of so many fields into these level basins brings to the three nations an enormous aggregate annual asset, a large proportion of which western nations are not yet utilizing. The greatest gain comes from the unfailing higher yields made possible by providing an abundance of water through which more plant food can be utilized, thus providing higher average yields. The waters used, coming as they do largely from the uncultivated hills and mountain lands, and carrying both dissolved and suspended matters, make positive annual additions of dissolved limestone and plant food elements to the fields; and in the aggregate they have been very large, through the persistent repetitions which have prevailed for centuries. If the yearly application of such water to the rice fields is but 16 inches, and if this water has the average composition quoted by Merrill for rivers of North America, taking into account neither suspended matter nor the absorption of potassium and phosphorus by it, each 10,000 square miles would receive, dissolved in the water, substances containing some 1,400 tons of phosphorus; 23,000 tons of potassium; 27,000 tons of nitrogen; and 48,000 tons of sulphur. In addition, there are brought to the fields some 216,000 tons of dissolved organic matter and a still larger weight of dissolved limestone, so necessary in neutralizing the acidity of soils, amounting to 1,221,000 tons. Such savings have been maintained in China, Korea and Japan on more than five, and possibly more than nine, times the 10,000 square miles, through centuries. The phosphorus thus turned upon 90,000 square miles would aggregate nearly 13,000,000 tons in a thousand years, which is less than the time the practice has been maintained, and the phosphorus is more than would be carried in the entire rock phosphate thus far mined in the United States, were it all 75 per cent pure.

The canalization of 50,000 square miles of our Gulf and Atlantic coastal plain, and the utilization on the fields of the silts and organic matter, together with the water, would mean turning to account a vast tonnage of plant food which is now wasting into the sea, and a correspondingly great increase of crop yield. There ought to be, and it would seem there must some time be, provided a way for sending to the sandy plains of Florida, and to the sandy lands between there and the Mississippi, large volumes of the rich silt and organic matter from this and other rivers, aside from that which should be applied systematically to building above flood plain the lands of the delta which are subject to overflow or are too low to permit adequate drainage.

It may appear to some that the application of such large volumes of water to fields, especially in countries of heavy rainfall, must result in great loss of plant food through leaching and surface drainage. But under the remarkable practices of these three nations this is certainly not the case, and it is highly important that we should understand and appreciate the principles which underlie the practices they have almost uniformly adopted on the areas devoted to rice irrigation. In the first place, their paddy fields are under-drained so that most of the water either leaves the soil through the crop, by surface evaporation, or it percolates through the subsoil into shallow drains. When water is passed directly from one rice paddy to another it is usually permitted some time after fertilization, when both soil and crop have had time to appropriate or fix the soluble plant food substances. Besides this, water is not turned upon the fields until the time for transplanting the rice, when the plants are already provided with a strong root system and are capable of at once appropriating any soluble plant food which may develop about their roots or be carried downward over them.

Although the drains are of the surface type and but 18 inches to 3 feet in depth, they are so numerous and close that, although the soil is nearly filled with water, there is a steady percolation of the fresh, fully aerated water carrying an abundance of oxygen into the soil to meet the needs of the roots. By this means it is made possible for watermelons, egg plants, musk melons and taro to be grown in rotation on the small paddy fields among the irrigated rice after the manner seen in the illustrations. In Fig. 136 each double row of egg plants is separated from the next by a narrow shallow trench which connects with a head drain and in which water is standing within 14 inches of the surface. The same was true in the case of the watermelons seen in Fig. 137, where the vines are growing on a thick layer of straw mulch which separates them from the moist soil, conserves the water by diminishing evaporation and, through decay from the summer rains and leaching, serves as fertilizer for the crop. In Fig. 138 the view is along a pathway separating two head ditches between areas in watermelons and taro, carrying the drainage waters from the several furrows into the main ditches. Although the soil appeared wet the plants were vigorous and healthy, seeming in no way to suffer from insufficient drainage.

These people have, therefore, given effective attention to the matter of drainage as well as irrigation and are looking after possible losses of plant food, as well as ways of supplying it. It is

FIG. 136. – Egg plants growing in the midst of rice fields with soil continually saturated and water standing in surface drain within 14 inches of the surface; Japan.

FIG. 137. – Watermelons, with the ground heavily mulched with straw, growing on low beds under conditions similar to those of Fig. 136.

not alone where rice is grown that cultural methods are made to conserve soluble plant food and to reduce its loss from the field, for very often, where flooding is not practised, small fields and beds, made quite level, are surrounded by low raised borders which permit not only the whole of any rain to be retained upon the field when so desired, but to be completely distributed over it, thus causing the whole soil to be uniformly charged with moisture

Fig. 138. – Looking along a path between two head ditches separating patches of watermelons and taro, Japan.

and preventing washing from one portion of the field to another. Such provisions are shown in Figs. 116 and 121.

Extensive as is the acreage of irrigated rice in China, Korea and Japan, nearly every spear is transplanted; the largest and best crop possible, rather than the least labour and trouble, determining their methods and practices. We first saw the fitting of the nursery rice beds at Canton and again near Kashing in Chekiang province on the farm of Mrs. Wu, whose homestead is seen in Fig. 139. She had come with her husband from Ningpo after the ravages of the Taiping rebellion had swept from two provinces alone

20,000,000 of people, and settled on a small area of then vacated land. As they prospered they added to their holding by purchase until about 25 acres were acquired, an area about ten times that possessed by the usual prosperous family in China. The widow was managing her place, one of her sons, although married, being still at school, the daughter-in-law living with her mother-in-law and helping in the home. Her field help during the summer consisted of seven labourers and she kept four cows for the ploughing and pumping of water for irrigation. The wages of the men were at the rate of $24, Mexican, for five summer months, together with

FIG. 139. – Residence compound and farm buildings of Mrs. Wu, Kashing, China.

their meals which were four each day. The cash outlay for the seven men was thus $14.45 of U.S. currency per month. Ten years before, such labour had been $30 per year, as compared with $50 at the time of our visit, or $12.90 and $21.50 of U.S. currency, respectively.

Her usual yields of rice were two piculs per mow, or $26\frac{2}{3}$ bushels per acre, and a wheat crop yielding half this amount, or some other crop, was taken from part of the land the same season, one fertilization answering for the two crops. She stated that her annual expense for fertilizers purchased was usually about $60, or $25.80 of U.S. currency. The homestead of Mrs. Wu, Fig. 139, consists of a compound in the form of a large quadrangle surrounding a court closed on the south by a solid wall 8 feet high. The structure is of earth brick, the roof of which is thatched with rice straw.

Our first visit here was April 19th. The nursery rice beds had been sown four days, at the rate of 20 bushels of seed per acre. The soil had been very carefully prepared and highly fertilized, the last treatment being a dressing of plant ashes so incompletely burned as to leave the surface coal black. The seed, scattered directly upon the surface, almost completely covered it and had been gently beaten into the dressing of ashes, with a wide, flat-bottom basket for the purpose. Each evening, if the night was likely to be cool, water was pumped over the bed, to be withdrawn the next day, if warm and sunny, so that the warmth might be absorbed by the black surface, and a fresh supply of air to be drawn into the soil.

FIG. 140. – Pumping station on the farm of Mrs. Wu, showing pump shelter, two power wheels connected with pumps, set at the end of a water channel leading from a canal.

Nearly a month later, May 14th, a second visit was made to this farm. and one of the nursery beds of rice, as it then appeared, is seen in Fig. 142; the plants were about 8 inches high and nearing the stage for transplanting. The field beyond the bed had already been partly flooded and ploughed, turning under 'Chinese clover' to ferment as green manure, preparatory for the rice transplanting. On the opposite side of the bed and in front of the residence, Fig. 139, flooding was in progress in the furrows between the ridges formed after the previous crop of rice was harvested and upon which the crop of clover for green manure was grown. At one end of the two series of nursery beds, one of which is seen in Fig. 142, was the pumping plant seen in Fig. 140, under a thatched shelter, with its two pumps installed at the end of a water channel leading from the canal. One of these wooden pump powers, with the blindfolded cow attached, is reproduced in Fig. 141.

More than a month is saved for maturing and harvesting winter and early spring crops, or in fitting the fields for rice, by this planting in nursery beds. The irrigation period for most of the land is cut short a like amount, saving both water and time. It is cheaper

FIG. 141.—Close view of power wheel with cow attached, used in driving the irrigation pump, one of the two seen in Fig. 140.

and easier to highly fertilize and prepare a small area for the nursery, while at the same time much stronger and more uniform plants are secured than would be possible by sowing in the field. The labour of weeding and caring for the plants in the nursery is far less than would be required in the field. It would be practically impossible to fit the entire rice areas as early in the season as the nursery beds are fitted, for the green manure is not yet grown and time is required for composting or for decaying, if ploughed under directly. The rice plants in the nursery are carried to a stage when they are strong feeders, and when set into the newly prepared, fertilized, clean soil of the field they are ready to feed strongly

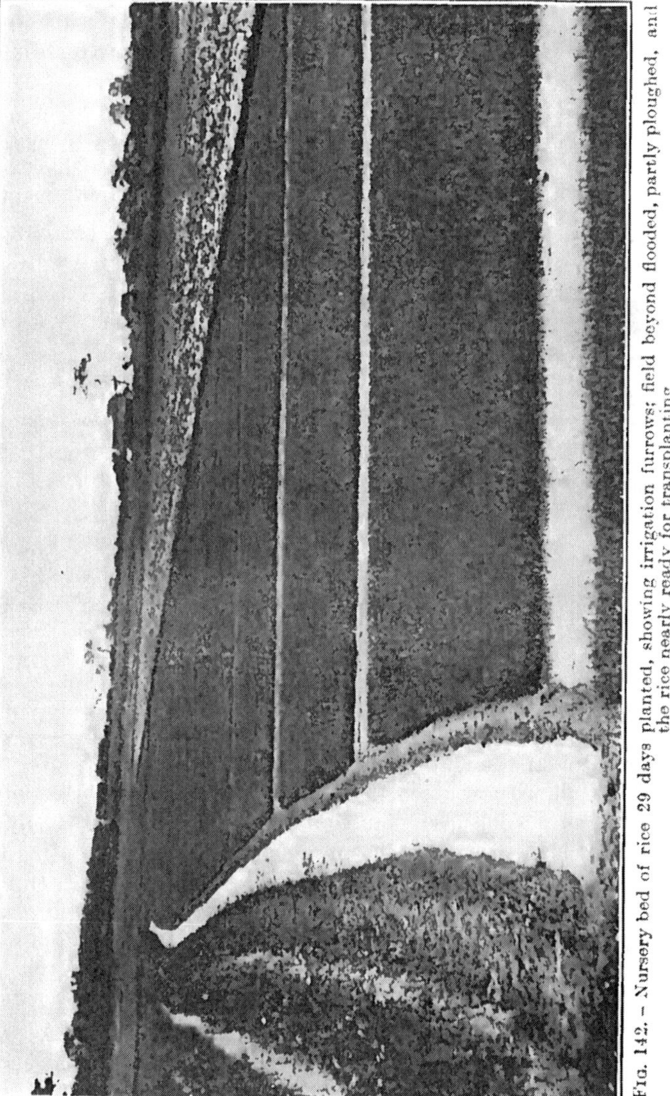

FIG. 142. – Nursery bed of rice 29 days planted, showing irrigation furrows; field beyond flooded, partly ploughed, and the rice nearly ready for transplanting

under the most favourable conditions. Both time and strength of plant are thus gained and these people are following what would appear to be the best possible practices under their condition of small holdings and dense population.

With our broad fields, our machinery and few people, their system appears to us crude and impossible, but cut our holdings to the size of theirs and our machinery, even our ploughs, would be impossible. So that the more one studies the environment of

Fig. 143. – Plough field nearly fitted for rice, and the smoothing, pulverizing harrow used for the purpose, Chekiang province.

these people, their numbers, what they have done and are doing, against what odds they have succeeded, the more difficult it is to see what course might have been better.

How full with work is the month which precedes the transplanting of rice has been pointed out – the making of the compost fertilizer; harvesting the wheat, rape and beans; distributing the compost over the fields, and their flooding and ploughing. In Fig. 143 one of these fields is seen ploughed, smoothed and nearly ready for the plants. The turned soil had been thoroughly pulverized, levelled and worked to the consistency of mortar, on the larger fields with one or another sort of harrow, as seen in Figs. 143 and 144. This thorough puddling of the soil permits the plants to be quickly set and provides conditions which ensure immediate perfect contact for the roots.

When the fields are ready women repair to the nurseries with their low four-legged bamboo stools, to pull the rice plants. Carefully rinsing the soil from the roots, they tie them into bundles of a size easily handled in transplanting, which are then distributed in the fields.

The work of transplanting may be done by groups labouring together after the manner seen in Fig. 146, made from four snapshots taken from the same point at intervals of fifteen minutes.

FIG. 144. – Form of revolving wooden harrow for fitting flooded rice fields preparatory to transplanting.

Long cords were stretched in the rice field 6 feet apart, and each of the seven men was setting six rows of rice 1 foot apart, six to eight plants in a hill, and the hills 8 or 9 inches apart in the row. The bundle was held in one hand and deftly, with the other, the desired number of plants were selected with the fingers at the roots, separated from the rest and, with a single thrust, set in place in the row. There was no packing of earth about the roots, each hill being set with a single motion, which followed one another in quick succession, completing one cross row of six hills after another. The men move backward across the field, completing one entire section, tossing the unused plants into the unset field; then reset the lines to cover another section. We were told that the usual day's work of transplanting, for a man under these conditions, after the field is fitted and the plants are brought to him, is two mow or one-third of an acre. The seven men in this group would thus set 2½ acres per day and, at the wage Mrs. Wu was paying, the cash outlay, if the help was hired, would be nearly 21 cents per acre. This is at a lower rate than we are able to set

Fig. 145. – Group of Chinese women pulling rice in a nursery bed, tying the plants in bundles preparatory to transplanting.

cabbage and tobacco plants with our best machine methods. In Japan, as seen in Figs. 147 and 148, the women participate in the work of setting the plants more than in China.

After the rice has been transplanted its care, unlike that of our wheat crop, does not cease. It must be hoed, fertilized and watered. To facilitate the watering all fields have been levelled, canals, ditches and drains provided. and, to aid in fertilizing and hoeing, the setting has been in rows and in hills in the row.

The first working of the rice fields after the transplanting, as we saw it in Japan, consisted in spading between the hills with a four-tined hoe, apparently more for loosening the soil and aeration than for killing weeds. After this treatment the field was gone

Fig. 146. – Transplanting rice in China. Four views taken from the same point at intervals of fifteen minutes, showing the progress made during forty-five minutes.

over again in the manner seen in Fig. 149, where the man is using his bare hands to smooth and level the stirred soil, taking care to eradicate every weed, burying them beneath the mud, and to straighten each hill of rice as it is passed. Sometimes the fingers are armed with bamboo claws to facilitate the weeding. Machinery in the form of revolving hand cultivators is now coming into use in Japan, and two men using these are seen in Fig. 12. In these cultivators the teeth are mounted on an axle so as to revolve as the cultivator is pushed along the row.

Fertilization for the rice crop receives the greatest attention

FIG. 147.—A group of Japanese women transplanting rice, in rainy weather costume, at Fukuoka Experiment Station.

everywhere, and in no direction more than in maintaining the store of organic matter in the soil. The pink clover, to which reference has been made (Figs. 84 and 85) is extensively sowed after a crop of rice is harvested in the autumn and comes into full bloom, ready to cut for compost or to turn under directly the rice fields are ploughed. Eighteen to twenty tons of this green clover are produced per acre, and in Japan this is usually applied to about 3 acres, the stubble and roots serving for the field producing the clover. Thus a dressing is given of 6 to 7 tons of green manure per acre, carrying not less than 37 pounds of potassium, 5 pounds of phosphorus, and 58 pounds of nitrogen.

Where the families are large and the holdings small, so that they

Fig. 148. — Japanese young women transplanting rice, under broad sunshade hats.

cannot spare room to grow the green manure crop, it is gathered on the mountain, weed and hill lands, or cut in the canals. On our boat trip west from Soochow the last of May, many boats were passed carrying tons of the long green ribbon-like grass, cut and gathered from the bottom of the canal. To cut this grass men were working to their armpits in the water of the canal, using a crescent-shaped knife mounted like an anchor from the end of a 16-foot bamboo handle. This was shoved forward along the bottom

FIG. 149. – Smoothing the soil and pulling weeds after the first working of a field of transplanted rice, Japan.

of the canal and then drawn backward, cutting the grass, which rose to the surface, where it was gathered on to the boats.

The straw of rice and other grain, the stems of any plant not usable as fuel and the chaff which is often scattered upon the water after the rice is transplanted (Fig. 151) may all be worked into the mud of rice fields.

Reference has been made to the utilization of waste of various kinds in these countries to maintain the productive power of their soils, but it is worth while, in the interests of western nations, as helping them to realize the ultimate necessity of such economies,

FIG. 150. — Boat-load of grass cut from bottom of canal, to be used as green manure or in preparing compost fertilizer, Kiangsu.

to state again, in more explicit terms, what Japan is doing. Dr. Kawaguchi, of the National Department of Agriculture and Commerce, taking his data from their records, informed me that Japan produced, in 1908, and applied to her fields, 23,850,295 tons of

FIG. 151. — Applying chaff to a rice field as a fertilizer.

human manure and 22,812,787 tons of compost; and that she imported 753,074 tons of commercial fertilizers, 7,000 of which were phosphates in one form or another. In addition to these she must have applied not less than 1,404,000 tons of fuel ashes and 10,185,500 tons of green manure products, grown on her hill and weed lands, and all these were applied to less than 14,000,000 acres of cultivated field. It should be emphasized that this is done because as yet she has found no better way of permanently maintaining a fertility capable of feeding her millions.

FIG. 152. – Well sweep and quadrangular conical water bucket used for irrigation in Chihli.

Besides fertilizing, transplanting and weeding the rice crop, there is the enormous task of irrigation to be maintained until the rice is nearly matured. Much of the water used is lifted by animal-power and a large share of it by man-power. The portable spool windlass, in Figs. 23 and 107, has been described, and Fig. 152 shows the quadrangular cone-shaped bucket and sweep extensively used in Chihli. This man was supplying water sufficient for the irrigation of half an acre per day, lifting the water 8 feet.

The form of pump most used in China and the foot-power for working it are seen in Fig. 153. Three men working a similar pump are seen in Fig. 133, a closer view of three men working the foot-power may be seen in Fig. 36, and still another stands adjacent to a series of flooded fields in Fig. 154. Where this view was taken the old farmer informed us that two men, with this pump, lifting water 3 feet, were able to cover two mow of land with 3 inches of water in two hours. This is at the rate of 2·5 acre-inches of water per ten hours per man, and for 12 to 15 cents, U.S. currency, thus making 16 acre-inches, or the season's supply of water, cost 77 to 96 cents, where coolie labour is hired and fed.

Fig. 153. – Three-man Chinese foot-power and wooden chain pump extensively used for irrigation in various parts of China.

Such is the efficiency of human power applied to the Chinese pump, measured in American currency.

This pump is simply an open box trough in which travels a wooden chain carrying a series of loosely-fitting boards which raise the water from the canal, discharging it into the field. The size of the trough and of the buckets are varied to suit the power applied and the amount of water to be lifted. Crude as it appears, there is nothing in western manufacture that can compete with it in first cost, maintenance or efficiency for Chinese conditions and nothing is more characteristic of all these people than their efficient, simple appliances of all kinds, which they have reduced to the lowest terms in every feature of construction and cost. The greatest results are accomplished by the simplest means. If a canal must be bridged, and it is too wide to be covered by a single span, the Chinese engineer may erect it at some convenient place and

FIG. 154.—Fields recently flooded with the Chinese foot power chain pump preparatory to ploughing for rice.

turn the canal under it when completed. This we saw in the case of a new railroad bridge near Sungkiang. The bridge was completed and the water had just been turned under it and compelled to make its own excavation. Great expense had been saved, while traffic on the canal had not been obstructed.

In the foot-power wheel of Japan all gearing is eliminated and the man walks the paddles themselves, as seen in Fig. 155. Some of these wheels are 10 feet in diameter, the diameter depending upon the height the water must be lifted.

Irrigation by animal power is extensively practised in each of the three countries, employing mostly the type of power wheel shown in Fig. 141. Fig. 156 shows the most common type of shelter seen in Chekiang and Kiangsu provinces, where they are very numerous. We counted as many as forty such shelters in a semicircle of half a mile radius. They provide comfort for the animals during both sunshine and rain, for under no conditions must the

Fig. 155. — Japanese irrigation foot-wheel.

water be permitted to run low on the rice fields, and everywhere domestic animals receive kind, thoughtful treatment.

In the less level sections, where streams have sufficient fall, current wheels are in common use, carrying buckets near their circumference arranged so as to fill when passing through the water, and to empty, after reaching the highest level, into a receptacle provided with a conduit which leads the water to the field. In Szechwan province some of these current wheels are so large and gracefully constructed as to strongly suggest Ferris wheels.

When the harvest time has come, notwithstanding the large

acreage of grain, yielding hundreds of millions of bushels, the small, widely-scattered holdings and the surface of the fields render all our machine methods impossible. Even our grain cradle, which preceded the reaper, would not do, and the great task is still met with the old-time sickle, as seen in Fig. 157, cutting the rice hill by hill, as it is transplanted.

Previous to the time for cutting, after the seed is well matured, the water is drawn off and the land permitted to dry and harden. The rainy season is not yet over and much care must be exercised

Fig. 156. – Power-wheel shelter on bank of canal, in Kiangsu province.

in curing the crop. The bundles may be shocked in rows along the margins of the paddy fields, as seen in Fig. 157, or they may be suspended, heads down, from bamboo poles, as seen in Fig. 158.

The threshing is accomplished by drawing the heads of the rice through the teeth of a metal comb mounted as seen at the right in Fig. 159, near the lower corner, behind the basket, where a man and woman are occupied in winnowing the dust and chaff from the grain by means of a large double fan. Fanning mills, built on the principle of those used by our farmers and closely resembling them, have long been used in both China and Japan. After the rice is threshed the grain must be hulled before it can

Fig. 157. – Japanese farmers harvesting rice with the old-time sickle.

Fig. 158. – Suspending rice bundles from bamboo frames set up in the fields for curing the grain, preparatory to threshing; Japan.

serve as food, and the oldest and simplest method of polishing used by the Japanese is seen in Fig. 160, where the friction of the grain upon itself does the polishing. A quantity of rice is poured into the receptacle when, with heavy blows, the long-headed plunger is driven into the mass of rice, thus forcing the kernels to slide over one another until, by their abrasion, the

Fig. 159. – Winnowing rice in Japan, using the large double fan worked by a pair of bamboo handles. A metal comb for removing the rice from the straw stands at the right.

desired result is secured. The same method of polishing, on a larger scale, is accomplished when the plungers are worked by the weight of the body, a series of men stepping upon lever handles of weighted plungers, raising them and allowing them to fall under the force of the weight attached. Recently, however, mills worked by petrol engines are in operation for both hulling and polishing, in Japan.

The many uses to which rice straw is put in the economies of these people make it almost as important as the rice itself. As food and bedding for cattle and horses; as thatching material for dwellings and other shelters; as fuel; as a mulch; as a source of organic matter in the soil, and as a fertilizer, it represents a money

Fig. 160. – Large wooden mortar used for the polishing of rice in Japan.

value which is very large. Besides these ultimate uses, the rice straw is extensively employed in the manufacture of articles used in enormous quantities. It is estimated that not less than 188,700,000 bags, such as are seen in Figs. 161 and 162, worth $3,110,000, are made annually from the rice straw in Japan. They are used for handling 346,150,000 bushels of cereals and 28,190,000 bushels of beans; and besides these, great numbers of bags are employed in transporting fish and other prepared manures.

In the prefecture of Hyogo, with 596 square miles of farm land, as compared with Rhode Island's 712 square miles, Hyogo farmers produced in 1906, on 265,040 acres, 10,584,000 bushels of rice worth $16,191,400, securing an average yield of almost 40 bushels per acre and a gross return of $61 for the grain alone. In addition to this, these farmers grew on the same land, the same

Fig. 161. – Sacking rice in bags made from the rice straw, Japan.

season, at least one other crop. Where this was barley the average yield exceeded 26 bushels per acre, worth $17.

In connection with their farm duties these Japanese families manufactured, from a portion of their rice straw, at night and during the leisure hours of winter, 8,980,000 pieces of matting and netting of different kinds having a market value of $262,000; 4,838,000 bags worth $185,000; 8,742,000 slippers worth $34,000; 6,254,000 sandals worth $30,000; and miscellaneous articles worth $64,000. This is a gross earning of more than $21,000,000 from eleven and a half townships of farm land and the labour of the farmers' families, an average earning of $80 per acre on nearly three-fourths of the farm land of this prefecture. At this rate three of the four forties of our 160-acre farms should bring a gross annual income of $9,600 and the fourth forty should pay the expenses.

At the Nara Experiment Station we were informed that the money value of a good crop of rice in that prefecture should be placed at $90 per acre for the grain and $8 for the unmanufactured straw; $36 per acre for the crop of naked barley, and $2 per acre

FIG. 162. – Loading, for shipment, rice put up in bags made from the rice straw; Japan.

for the straw. The farmers here practise a rotation of rice and barley covering four or five years, followed by a summer crop of melons, worth $320 per acre, and some other vegetable instead of the rice on the fifth or sixth year, worth 80 yen per tan, or $160 per acre. To secure green manure for fertilizing, soy beans are planted each year in the space between the rows of barley, the barley being planted in November. One week after the barley is harvested the soy beans, which produce a yield of 160 kan per tan, or 5,290 pounds per acre, are turned under and the ground fitted for rice. At these rates the Nara farmers are producing on four-fifths or five-sixths of their rice lands a gross earning of $136 per acre annually, and on the other fifth or sixth an earning of $480 per acre, not counting the annual crop of soy beans used in maintaining the nitrogen and organic matter in their soils, and not counting their earnings from home manufactures. Can the farmers of our South Atlantic and Gulf Coast states, which are in the same latitude, some time attain to this standard? We see no reason why they should not, but only with the best of irrigation and fertilization, with proper rotation and multiple cropping.

XIII

SILK CULTURE

In some ways one of the most remarkable industries of the Orient is that of silk production, and its manufacture into the most exquisite and beautiful fabrics in the world. Remarkable for its magnitude; for having had its birthplace apparently in oldest China, at least 2,600 years B.C.; for having been founded on the domestication of a wild insect of the woods; and for having lived through more than 4,000 years, and expanded until to-day a $1,000,000 cargo of the product is laid down on our western coast at one time and rushed by special express to New York City for the Christmas trade.

Japan produced in 1907 26,072,000 pounds of raw silk from 17,154,070 bushels of cocoons, feeding the silkworms from mulberry leaves grown on 957,560 acres. At the export selling price of this silk in Japan the crop represents a money value of $124,000,000, or more than $2 *per capita* for the entire population of the Empire; and engaged in the care of the silkworms, illustrated in Figs. 163, 164, 165 and 166, there were, in 1906, 1,407,766 families, or some 7,000,000 people.

Richard's *Geography of the Chinese Empire* places the total export of raw silk to all countries, from China, in 1905, at 30,413,200 pounds, and this, at the Japanese export price, represents a value of $145,000,000. Richard also states that the value of the annual Chinese export of silk to France amounts to 10,000,000 pounds sterling, and that this is but 12 per cent of the total, from which it appears that her total export alone reaches a value near $400,000,000.

The use of silk in wearing apparel is more general among the Chinese than among the Japanese, and with China's eightfold greater population, the home consumption of silk must be large indeed and her annual production must far exceed that of Japan. Hosie places the output of raw silk in Szechwan at 5,439,500 pounds, which is nearly a quarter of the total output of Japan, and silk is extensively grown in eight other provinces, which together have an area nearly fivefold that of Japan. It would appear, therefore, that a low estimate of China's annual production of raw silk must be some 120,000,000 pounds, and this, with the output of Japan and Korea, would make a product for the three countries probably exceeding 150,000,000 pounds annually, representing a total value of perhaps $700,000,000; quite equalling in value the wheat crop of the United States, but produced on less than one-eighth of the area.

According to the observations of Count Dandola, the worms

FIG. 163. – Removing silkworm eggs from sheets of paper where they were laid preparatory to hatching, Japan.

which contribute to this vast earning are so small at hatching that some 700,000 of them weigh only one pound; but they grow very rapidly, shed their skins four times, weighing 15 pounds at the time of the first moult, 94 pounds at the second, 400 pounds at the third, 1,628 pounds at the fourth moulting, and when mature have come to weigh nearly 5 tons – 9,500 pounds. In making this growth during about thirty-six days, according to Paton, the 700,000 worms have eaten 105 pounds by the time of the first moult; 315 pounds by the second; 1,050 pounds by the third; 3,150 pounds by the fourth; and in the final period, before spinning,

19,215 pounds, thus consuming in all nearly 12 tons of mulberry leaves in producing nearly 5 tons of live weight, or at the rate of $2\frac{1}{2}$ pounds of green leaf to 1 pound of growth.

According to Paton, the cocoons from the 700,000 worms would weigh between 1,400 and 2,100 pounds, and these, according to the observations of Hosie in the province of Szechwan, would yield about one-twelfth their weight of raw silk. On this basis the 1 pound of worms hatched from the eggs would yield between

FIG. 164. – Feeding silkworms. One of the 16 bamboo trays, on which the silkworms are feeding, has been removed from the racks and Japanese girls are spreading over it a fresh supply of mulberry leaves.

116 and 175 pounds of raw silk, worth, at the Japanese export price for 1907, between $550 and $832, and 164 pounds of green mulberry leaves would be required to produce a pound of silk.

A Chinese banker in Chekiang province, with whom we talked, stated that the young worms which would hatch from the eggs spread on a sheet of paper 12 by 18 inches would consume, in coming to maturity, 2,660 pounds of mulberry leaves and would spin 21·6 pounds of silk. This is at the rate of 123 pounds of leaves to 1 pound of silk. The Japanese crop for 1907, 26,072,000 pounds, produced on 957,560 acres, is a mean yield of 27·23 pounds of raw silk per acre of mulberries, and this would require a mean yield of 4,465 pounds of green mulberry leaves per acre, at the rate of 164 pounds per pound of silk.

Ordinary silk in these countries is produced largely from three varieties of mulberries, and from them there may be three pickings of leaves for the rearing of a spring, summer and autumn crop of silk. We learned at the Nagoya Experiment Station, Japan, that

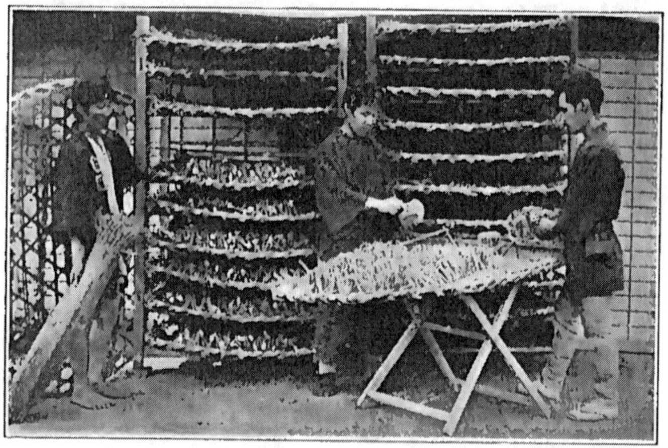

Fig. 165. – Providing places for silkworms to spin their cocoons.

there good spring yields of mulberry leaves are at the rate of 400 kan, the second crop 150 kan, and the third crop 250 kan per tan, making a total yield of over 13 tons of green leaves per acre. This, however, seems to be materially higher than the average for the Empire.

In Fig. 167 is a near view of a mulberry orchard in Chekiang province, which had been very heavily fertilized with canal mud, and which was at the stage for cutting the leaves to feed the first crop of silkworms. A bundle of cut limbs is in the crotch of the front tree in the view. Those who raise mulberry leaves are not usually the feeders of the silkworms, and the leaves from this

FIG. 166. – Selecting the best cocoons, male and female determined by the shape and size, for purposes of breeding.

orchard were being sold at one dollar, Mexican, per picul, or 32·25 cents per 100 pounds. The same price was being paid a week later in the vicinity of Nanking, Kiangsu province.

The mulberry trees, as they appear before coming into leaf in

FIG. 167. – A near view of a mulberry orchard in Chekiang province.

the early spring, may be seen in Fig. 168. The long limbs are the shoots of the last year's growth, from which at least one crop of

Fig. 168. – Near view of mulberry tree many years old, showing limbs of the last year's growth which will be cut close to the old wood when in full leaf.

leaves had been picked, and in healthy orchards they may have a length of 2 to 3 feet. An orchard, from a portion of which the limbs had just been cut, presented the appearance seen in Fig. 169. These trees were twelve to fifteen years old and the enlargements on the ends of the limbs resulted from the frequent pruning, year after year, at nearly the same place. The ground under these trees was thickly covered with a growth of pink clover just coming into bloom, which would be spaded into the soil, providing nitrogen and organic matter, whose decay would liberate potash, phosphorus and other mineral plant-food elements for the crop.

In Fig. 170 three rows of mulberry trees, planted 4 feet apart, stand on a narrow embankment raised 4 feet, partly through adjusting the surrounding fields for rice, and partly by additions of canal mud used as a fertilizer. On either side of the mulberries is a crop of windsor beans, and on the left a crop of rape, both of which would be harvested in early June, the ground where they stand flooded, ploughed and transplanted to rice. This and the other mulberry views were taken in the extensively canalized portion of China represented in Fig. 45. The farmer owning this orchard had just finished cutting two large bundles of limbs for the sale of the leaves in the village. He stated that his first crop ordinarily yields from 3 to as many as 20 piculs per mow, but that the second crop seldom exceeded 2 to 3 piculs. The first and second crop of leaves, if yielding together 23 piculs per mow, would amount to 9·2 tons per acre, worth, at the price named, $59.34. Mulberry leaves must be delivered fresh as soon as gathered and must be fed the same day; the limbs, when stripped of their leaves, at the place where these are sold, are tied into bundles and reserved for use as fuel.

In the south of China the mulberry is grown from low cuttings rooted by layering. We have before spoken of our five hours' ride in the Canton delta region, on the steamer *Nanning,* through extensive fields of low mulberry, then in full leaf, which were first mistaken for cotton nearing the blossom stage. This form of mulberry is seen in Fig. 37, and the same method of pruning is practised in southern Japan. In middle Japan high pruning, as in Chekiang and Kiangsu provinces, is followed, but in northern Japan the leaves are picked directly, as is the case everywhere with the last crop of leaves. Pruning is not practised in the more northern latitudes.

Not all silk produced in these northern countries is from the

FIG. 169. – Mulberry orchard recently pruned for the first crop of leaves, with unpruned trees along the right.

Fig. 170. – Three rows of mulberry trees occupying a long narrow embankment, which will be surrounded later by flooded rice fields.

domesticated *Bombyx mori*; large amounts are obtained from the spinnings of wild silkworms feeding upon the leaves of a species of oak growing on the mountain and hill lands in various parts of China, Korea and Japan. In China the collections in largest amount are reeled from the cocoons of the tussur worm (*Antheræa pernyi*) gathered in Shantung, Honan, Kweichow and Szechwan provinces. In the hilly parts of Manchuria also this industry is attaining large proportions, the cocoons being sent to Chefoo in the Shantung province, to be woven into pongee silk.

M. Randot has estimated the annual crop of wild silk cocoons in Szechwan at 10,180,000 pounds, although in the opinion of Alexander Hosie much of this may come from Kweichow. Richard places the export of raw wild silk from the whole of China proper, in 1904, at 4,400,000 pounds. This would mean not less than 75,300,000 pounds of wild cocoons and may be less than half the home consumption.

From data collected by Alexander Hosie, it appears that in 1899 the export of raw tussur silk from Manchuria, through the port of Newchwang by steamer alone, was 1,862,448 pounds, valued at $1,721,200, and the production is increasing rapidly. The export from the same port the previous year, by steamer, was 1,046,704 pounds. This all comes from the hilly and mountain lands south of Mukden, lying between the Liao plain on the west and the Yalu river on the east, covering some 5,000 square miles, which we crossed on the Antung-Mukden railway.

There are two broods of these wild silkworms each season, between early May and early October. Cocoons of the autumn brood are kept through the winter. When the moths come forth they are caused to lay their eggs on pieces of cloth, and when the worms are hatched they are fed until the first moult upon the succulent new oak leaves gathered from the hills, after which the worms are taken to the low oak growth on the hills where they feed themselves and spin their cocoons under the cover of leaves drawn about them.

The moths reserved from the first brood, after becoming fertile, are tied by means of threads to the oak bushes, where they deposit the eggs which produce the second crop of tussur silk. To maintain an abundance of succulent leaves within reach the oaks are periodically cut back.

Thus these plain people, patient, frugal, unshrinking from toil, the basic units of three of the oldest nations, go to the uncultivated hill lands and from the wild oak and the millions of insects which they help to feed upon it, not only create a valuable export trade, but procure material for clothing, fuel, fertilizer and food, for the large chrysalides, cooked in the reeling of the silk, may be eaten at once or are seasoned with sauce to be used later. Besides this, the last unreelable portion of each cocoon is laid aside to be manufactured into silk wadding and soft mattresses for caskets upon which the wealthy lay their dead.

XIV

THE TEA INDUSTRY

THE cultivation of tea in China and Japan is another of the great industries of these nations, taking rank with that of sericulture, if not above it, in the important part it plays in the welfare of the people. There is little reason to doubt that the industry has its foundation in the need of something to render boiled water palatable for drinking purposes. The drinking of boiled water has been universally adopted in these countries as an individually available and thoroughly efficient safeguard against that class of deadly disease germs which it has been almost impossible to exclude from the drinking water of any densely peopled country.

So far as may be judged from the success of the most thorough sanitary measures thus far instituted, and taking into consideration the inherent difficulties which must increase enormously with increasing populations, it appears inevitable that modern methods must ultimately fail in sanitary efficiency and that absolute safety must be secured in some manner having the equivalent effect of boiling water, the method which was long ago adopted by the Mongolian races, and which destroys active disease germs at the latest moment before using. It must not be overlooked that the boiling of drinking water in China and Japan has been demanded quite as much because of congested rural populations as to guard against such dangers in large cities, while as yet our sanitary engineers have dealt only with the urban phases of this most vital problem and chiefly, too, thus far, only where it has been possible to procure the water supply in comparatively unpopulated hill lands. But such opportunities cannot remain available indefinitely, any more than they did in China and Japan, and already typhoid epidemics break out in our large cities and citizens are advised to boil their drinking water.

If tea-drinking in the family is to remain general in most portions of the world, and especially if it increases in proportion to population, there is great industrial and commercial promise for China, Korea and Japan in their tea industry; particularly if they develop tea culture still further over the extensive and still unused flanks of the hill lands, improve their cultural methods, their manufacture, and develop their export trade. They have the best of climatic and soil conditions and people sufficiently capable of enormously expanding the industry. Both improvement and expansion of methods along all essential lines are needed, enabling them to put upon the market pure teas of thoroughly uniform grades of guaranteed quality, and with these the maintenance of an international code of rigid ethics which will secure to all concerned a square deal and a fair division of the profits.

The production of rice, silk and tea are three industries which these nations are pre-eminently circumstanced and qualified to economically develop and maintain. Other nations may better specialize along other lines, and the time is coming when maximum production at minimum cost, as the result of clean robust living, will determine lines of social progress and of international relations. With the vital awakening to the possibility of and necessity for world peace, it must be recognized that this can be nothing less than universal, industrial, commercial, intellectual and religious, in addition to making impossible for ever the carnage that has ravaged the world through all the centuries.

With the extension of rapid transportation and more rapid communication throughout the world, we are fast entering the state of social development which will treat the whole world as a harmonious industrial unit. It must be recognized that in certain regions, because of peculiar fitness of soil, climate and people, needful products can be produced there better and so much more cheaply than elsewhere as to pay the cost of transportation. If China, Korea and Japan, with parts of India, can and will produce the best and cheapest silks, teas or rice, it must be for the greatest good to seek a mutually helpful exchange, whereas the erection of impassable tariff barriers is a declaration of war and cannot make for world peace and world progress.

The date of the introduction of tea culture into China seems to be unknown. It was before the beginning of the Christian era, and tradition would place it more than 2,700 years earlier. The Japanese definitely date its introduction into their islands as in the year A.D. 805, and state its coming to them from China. However and whenever tea-growing originated in these countries, it long ago attained, and now maintains, large proportions. In 1907 Japan had 124,482 acres of land occupied by tea gardens and tea plantations. These produced 60,877,975 pounds of cured tea, giving a mean yield of 489 pounds per acre. Of the more than 60,000,000 pounds of tea produced annually on nearly 200 square miles in Japan, less than 22,000,000 pounds are consumed at home, the balance being exported at a cash value, in 1907, of $6,309,122, or a mean of 16 cents per pound.

In China the volume of tea produced annually is much larger than in Japan. Hosie places the annual export from Szechwan into Tibet alone at 40,000,000 pounds, and this is produced largely

FIG. 171. – Near view of tea garden with ground heavily mulched with straw, adjoining a Japanese farm village.

in the mountainous portion of the province west of the Min river. Richard places her direct export to foreign countries, in 1905, at 176,027,255 pounds; and in 1906 at 180,271,000 pounds, so that the annual export must exceed 200,000,000 pounds, and her total product of cured tea must be more than 400,000,000 pounds.

The general appearance of tea bushes as they are grown in Japan is indicated in Fig. 171. The form of the bushes, the shape and size of the leaves and the dense green, shiny foliage quite suggests our box, so much used in borders and hedges. When the bushes are young, not covering the ground, other crops are grown between the rows, but as the bushes attain their full size, standing after trimming, waist to breast high, the ground between is usually thickly covered with straw, leaves or grass, and weeds from the hill lands, which serve as a mulch, as a fertilizer, as a means of preventing washing on the hill-sides, and to force the rain to enter the soil uniformly where it falls.

Quite a large percentage of the tea bushes are grown on small, scattered, irregular areas about dwellings, or on land not readily

FIG. 172. – Looking across a tea plantation located on the flanks of wooded hill lands rising in the background, Japan.

tilled, but there are also many tea plantations of considerable size, presenting the appearance seen in Fig. 172. After each picking of the leaves the bushes are trimmed back with pruning-shears, giving the rows the appearance of carefully trimmed hedges.

The tea-leaves are hand-picked, generally by women and girls, after the manner seen in Fig. 173, where they are gathering the tender newly-formed leaves into baskets to be weighed fresh, as seen in Fig. 174.

Three crops of leaves are usually gathered each season, the first yielding in Japan 100 kan per tan, the second 50 kan, and the third 80 kan per tan. This is at the rate of 3,307 pounds, 1,653 pounds, and 2,645 pounds per acre, making a total of 7,605 pounds for the season, from which the grower realizes from a little more than 2·2 to a little more than 3 cents per pound of the green leaves, or a gross earning of $167 to $209.50 per acre.

Fig. 173. – Group of Japanese women picking leaves of the tea plant.

We were informed that the usual cost for fertilizers for the tea orchards was 15 to 20 yen per tan, or $30 to $40 per acre per annum, the fertilizer being applied in the autumn, in the early spring, and again after the first picking of the leaves. While the tea plants are yet small one winter crop and one summer crop of vegetables, beans or barley are grown between the rows, these giving a return of some $40 per acre. Where the plantations are given good care and ample fertilization the life of a plantation may be prolonged continuously, it is said, through one hundred or more years.

During our walk from Joji to Kowata, along a country road in

FIG. 174. – Weighing the freshly picked tea-leaves in Japan.

one of the tea districts, we passed a tea-curing house. This was a long rectangular one-story building with twenty furnaces arranged, each under an open window, around the sides. In front of each heated furnace with its tray of leaves, a Japanese man, wearing only a breech cloth, and in a state of profuse perspiration, was busy rolling the tea leaves between the palms of his hands.

At another place we witnessed the making of the low-grade dust tea, which is prepared from the leaves of bushes which must be removed or from those of the prunings. In this case the dried bushes with their leaves were being beaten with flails on a threshing floor. The dust tea thus produced is consumed by the poorer people.

XV

ABOUT TIENTSIN

On the 6th of June we left central China for Tientsin and further north, sailing by coastwise steamer from Shanghai, again ploughing through the turbid waters which give literal exactness to the name Yellow Sea. Our steamer touched at Tsingtao, taking on board a body of German troops, and again at Chefoo, and it was only between these two points that the sea was not strongly turbid. Nor was this all. From early morning of the 10th until we anchored at Tientsin, 2.30 p.m., our course up the winding Pei-ho was against a strong dust-laden wind, which left those who had kept to the deck as grey as though they had ridden by automobile through the Colorado desert. So the soils of high interior Asia are still spreading eastward by flood and wind into the valleys and far over the coastal plains. Over large areas between Tientsin and Peking, and at other points northward toward Mukden, trees and shrubs have been systematically planted in rectangular hedgerow lines, to check the force of the winds and reduce the drifting of soils, planted fields occupying the spaces between.

It was on this trip that we met Dr. Evans of Shunking, Szechwan province. His wife is a physician practising among the Chinese women, and in discussing the probable rate of increase of population among the Chinese, it was stated that she had learned through her practice that very many mothers had borne seven to eleven children, and yet but one, two, or at most three, were living. It was said there are many customs and practices which determine this high mortality among children, one of which is that of feeding them meat before they have teeth, the mother masticating for the children, with the result that often fatal convulsions follow. A Scotch physician of long experience in Shantung, who took the steamer at Tsingtao, replied to my question as to the usual size of families in his circuit, 'I do not know. It depends on the crops. In good years the number is large; in times of famine the girls especially are disposed of, often permitted to die when very young for lack of care. Many are sold at such times to go into other provinces.' Such statements, however, should doubtless be taken with much allowance. If all the details were known regarding the cases which have served as foundations for such reports, the matter might appear in quite a different light from that suggested by such cold recitals.

Although land-taxes are high in China, Dr. Evans informed me that it is not infrequent for the same tax to be levied twice and even three times in one year. Inquiries regarding the land-taxes among farmers in different parts of China showed rates running from 3 cents to $1½, Mexican, per mow; or from about 8 cents to $3.87, gold, per acre. At these rates a 40-acre farm would pay from $3.20 to $154.80, and a quarter section four times these amounts. Data collected by Consul-General E. T. Williams of Tientsin indicate that in Shantung the land-tax is about $1 per acre, and in Chihli 20 cents. In Kiangsi province the rate is 200 to 300 cash per mow, and in Kiangsu, from 500 to 600 cash per mow, or, according to the rate of exchange given on page 76, from 60 to 80 cents, or 90 cents to $1.20 per acre in Kiangsi; and $1.50 to $2.00 or $1.80 to $2.40 in Kiangsu province. The lowest of these rates would make the land-tax on 160 acres $96, and the highest would place it at $384, gold.

In Japan the taxes are paid quarterly and the combined amount of the national, prefectural and village assessments usually aggregates about 10 per cent of the government valuation placed on the land. The mean valuation placed on the irrigated fields, excluding Formosa and Karafuto, was in 1907, 35·35 yen per tan; that of the upland fields, 9·40 yen, and the *genya* and pasture lands were given a valuation of ·22 yen per tan. These are valuations of $70.70, $18.80 and $.44, gold, per acre, respectively, and the taxes on 40 acres of paddy field would be $282.80; $75.20 on 40 acres of upland field, and $1.76, gold, on the same area of the *genya* and weed lands.

In the villages, where work of one kind or another is done for pay, Dr. Evans stated that a woman's wage might not exceed $8, Mexican, or $3.44, gold, per year, and when we asked how it could be worth a woman's while to work a whole year for so small a sum, his reply was, 'If she did not do this she would earn nothing, and this would keep her in clothes and a little more.' A cotton-spinner in his church would procure a pound of cotton and on returning the yarn would receive 1¼ pounds of cotton in exchange, the quarter pound being her compensation.

Dr. Evans also described a method of rooting slips from trees, practised in various parts of China. The under side of a branch is cut, bent upward and split for a short distance; about this is packed a ball of moistened earth wrapped in straw to retain the soil and to provide for future watering; the whole may then be bound with strips of bamboo for greater stability. In this way slips for new mulberry orchards are procured.

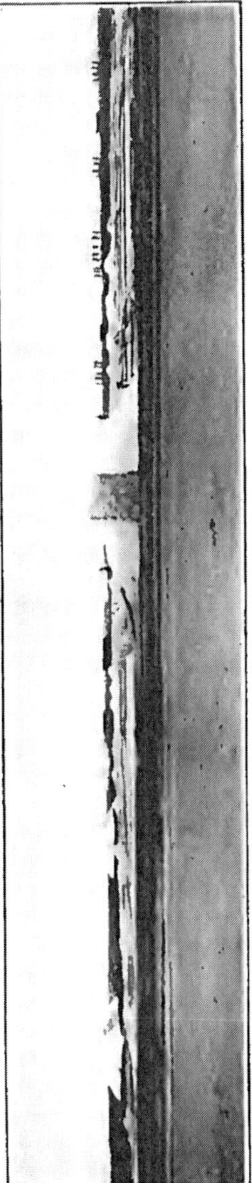

Fig. 175. – Salt stacks and sail windmills on the Taku evaporation fields at the mouth of the Pei-ho, Chihli.

At eight o'clock in the morning we entered the mouth of the Pei-ho and wound westward through a vast, nearly sea-level, desert plain. In both directions, far toward the horizon, huge white stacks of salt dotted the surface of the Taku Government salt fields, and revolving in the wind were great numbers of horizontal sail windmills, pumping sea-water into an enormous acreage of evaporation basins. In Fig. 175 may be seen five of the large salt stacks and six of the windmills, together with many smaller piles of salt. Fig. 176 is a closer view of the evaporation basins with piles of salt scraped from the surface after the mother liquor has been drained away. The windmills, which were working one, sometimes two, of the large wooden chain pumps, were some 30 feet in diameter and lifted the brine from tide-water basins into those of a second and third higher level where the second and final concentration occurred. These windmills, crude as they appear in Fig. 177, are nevertheless efficient, cheaply constructed and easily controlled. The eight sails, each 6 by 10 feet, were so hung as to take the wind through the entire revolution, tilting automatically to receive the wind on the opposite face the moment the edge passed the critical point. Some 480 feet of sail surface were thus spread to the wind, working on a radius of 15 feet. The horizontal drive wheel had a diameter of 10 feet, and carried eighty-eight wooden cogs which engaged a pinion with fifteen leaves. There were nine

FIG. 176. – Near view of evaporating basins with piles of salt ready to be removed from the fields.

arms on the reel at the other end of the shaft which drove the chain. The boards or buckets of the chain pump were 6 by 12 inches, placed 9 inches apart, and with a fair breeze the pump ran full.

Enormous quantities of salt are thus cheaply manufactured by wind, tide and sun power directed by the cheapest human labour. Before reaching Tientsin we passed the Government storage yards and counted 200 stacks of salt piled in the open, and more than a third of the yard had been passed before beginning the count.

The average content of each stack must have exceeded 3,000 cubic feet of salt, and more than 40,000,000 pounds must have been stored in the yards. Armed guards in military uniform patrolled the alleyways day and night. Long strips of matting laid over the stacks were the only shelter against rain.

Throughout the length of China's sea-coast, from as far north

Fig. 177. - Sail windmill used in pumping brine at the Taku Government salt works, Chihli.

as beyond Shanhaikwan, south to Canton, salt is manufactured from sea-water in suitable places. In Szechwan province, we learn from the report of Consul-General Hosie that not less than 300,000 tons of salt are annually manufactured there, largely from brine raised by animal power from wells 700 to more than 2,000 feet deep.

Hosie describes the operations at a well more than 2,000 feet deep, at Tzelintsing. In the basement of a power-house which sheltered forty water buffaloes, a huge bamboo drum 12 feet high, 60 feet in circumference, was so set as to revolve on a vertical axis propelled by four cattle drawing from its circumference. A hemp rope was wound about this drum, 6 feet from the ground, passing out and under a pulley at the well, then up and around a wheel mounted 60 feet above and descending to a bucket, made from bamboo stems, which dropped with great speed to the bottom of the well as the rope unwound. When the bucket reached the bottom four attendants, each with a buffalo in readiness, hitched to the drum and drove at a running pace during fifteen minutes, or until the bucket was raised from the well. The buffalo were then unhitched and, while the bucket was being emptied and again dropped to the bottom of the well, a fresh relay were brought to the drum. In this way the work continued night and day.

The brine, after being raised from the well, was emptied into distributing reservoirs, flowing thence through bamboo pipes to the evaporating sheds, where round-bottomed, shallow iron kettles 4 feet across were set in brick arches in which jets of natural gas were burning.

Within an area some 60 miles square there are more than a thousand brine and twenty fire wells from which fuel gas is taken. The mouths of the fire wells are closed with masonry, out from which bamboo conduits coated with lime lead to the various furnaces, terminating with iron burners beneath the kettles. Remarkable is the fact that in the city of Tzeliutsing both these brine and fire wells have been operated in the manufacture of salt since before Christ was born.

The forty water buffalo are worth $30 to $40 per head and their food 15 to 20 cents per day. The cost of manufacturing the salt is placed at 13 to 14 cash per catty, to which the Government adds a tax of 9 cash more, making the cost at the factory from 82 cents to $1.15, gold, per 100 pounds. Salt manufacture is a Government monopoly and the product must be sold either to Government officials or to merchants who have bought the exclusive right to supply certain districts. The importation of salt is prohibited by treaties. For the salt tax collection China is divided into eleven circuits, each having its own source of supply, and transfer of salt from one circuit to another is forbidden.

The usual cost of salt is said to vary between one and a half and four cash per catty. The retail price of salt ranges from three-fourths to three cents per pound, fully twelve to fifteen times the cost of manufacture. The annual production of salt in the Empire is some 1,860,000 tons, and in 1901 salt paid a tax close to 10,000,000 dollars.

Beyond the salt fields, toward Tientsin, the banks of the river were dotted at short intervals with groups of low, almost windowless houses (Fig. 178) built of earth brick plastered with clay on

Fig. 178. – Chinese village on the bank of the Pei-ho, Province of Chihli.

sides and roof, made more resistant to rain by an admixture of chaff and cut straw. There was a remarkable freshness of look about them which we learned was the result of recent preparations made for the rainy season about to open. Beyond the first of these villages came a stretch of plain dotted thickly with innumerable grave mounds, to which reference has been made. For nearly an hour we had travelled up the river before there was any material vegetation, the soil being too saline apparently to permit growth, but beyond this, crops in the fields and gardens, with some fruit and other trees, formed a fringe of varying width along the banks. Small fields of transplanted rice on both banks were frequent and often the land was laid out in beds of two levels, carefully graded, the rice occupying the lower areas, and wooden chain pumps were being worked by hand, foot and animal power, irrigating both rice and garden crops.

In the villages were many stacks of earth compost, of the Shantung type; manure middens were common and donkeys drawing heavy stone rollers, followed by men with large wooden mallets, were going round and round, pulverizing and mixing the dry earth compost and the large earthen brick from dismantled kangs, preparing fertilizer for the new series of crops about to be planted Large boatloads of these prepared fertilizers were moving on the river and up the canals to the fields.

Toward the coast from Tientsin, especially in the country traversed by the railroad, there was little produced except a short grass, this being grazed at the time of our visit and, in places, cut for a very meagre crop of hay. The productive cultivated lands lie chiefly along the rivers and canals or other watercourses, where there is better drainage as well as water for irrigation. The extensive, close canalization that characterizes parts of Kiangsu and Chekiang provinces is lacking here, and for this reason, in part, the soil is not so productive. The fuller canalization, the securing of adequate drainage and the gaining of complete control of the flood waters which flow through this vast plain during the rainy season, constitute one of China's most important industrial problems, which, when properly solved, must vastly increase her resources. During our drive over the old Peking-Taku road saline deposits were frequently observed which had been brought to the surface during the dry season, and the city engineer of Tientsin stated that in their efforts at parking portions of the foreign concessions they had found the trees dying after a few years when their roots began to penetrate the more saline subsoil, but that since they had opened canals, improving the drainage, trees were no longer dying. There is little doubt that proper drainage by means of canals, and the irrigation which would go with it, would make all of these lands, now more or less saline, highly productive, as are now those contiguous to the existing watercourses.

It had rained two days before our drive over the Taku road, and when we applied for a conveyance the proprietor doubted whether the roads were passable, as he had been compelled to send out an extra team to assist in the return of one which had been stalled during the previous night. It was finally arranged to send an extra horse with us. The rainy season had just begun and the deep trenching of the roads concentrates the water in them and greatly intensifies the trouble. In one of the little hamlets through which we passed the roadway was trenched to a depth of 3 to 4 feet in the middle of the narrow street, leaving only 5 feet for passing in front of the dwellings on either side, and in this

Fig. 179. – China's method of shallow cultivation, producing an earth mulch to conserve soil moisture.

trench our carriage moved through mud and water nearly to the hubs.

Between Tientsin and Peking, in the early morning after a rain of the night before, we saw many farmers working their fields with the broad hoes, developing an earth mulch at the first possible moment to conserve their much-needed moisture. Men were at work, as seen in Figs. 179 and 180, using long-handled hoes, with blades 9 by 13 inches, hung so as to draw just under the surface. They were very effective and permitted the men to cover the ground rapidly.

Walking farther, we came upon six women in a field of wheat, gleaning the single heads which had prematurely ripened and broken over upon the ground between the rows soon to be harvested. Whether they were doing this as a privilege or as a task we do not know; they were strong, cheerful, reasonably dressed, hardly past middle life, and it was nearly noon, yet not one of them had collected more straws than she could readily grasp in one hand. The season in Chihli, as in Shantung, had been one of unusual drought, making the crop short, and perhaps unusual frugality was being practised; but it is in saving that these people

FIG. 180. – Hoe used for shallow cultivation in developing an earth mulch. The blade is 13 inches long and 9 inches wide.

excel perhaps more even than in producing. The heads of wheat, if left upon the ground, would be wasted, and if the women were privileged gleaners in the fields their returns were certainly much greater than were those of the very old women we have seen in France gathering heads of wheat from the already harvested fields.

In the fields between Tientsin and Peking all wheat was being pulled, the earth shaken from the roots, tied in small bundles and taken to the dwellings, sometimes on the heavy cart drawn by a team consisting of a small donkey and cow hitched tandem, as seen in Fig. 181. Millet had been planted between the rows of wheat in this field and was already up. When the wheat was removed the ground would be fertilized and planted with soy beans. Because of the dry season this farmer estimated his yield would be but 8 to 9 bushels per acre. He was expecting to harvest 13 to 14 bushels of millet and from 10 to 12 bushels of soy beans per acre from the same field. This would give him an earning, based on the local prices, of $10.36, gold, for the wheat, $6.00 for

Fig. 181. – Gathering wheat, harvested by being pulled and tied in bundles. Team consists of a small donkey and a medium-sized cow, which constitute the most common farm team. Tientsin.

the beans, and $5.48 per acre for the millet. The land was owned by the family of the Emperor and was rented at $1.55, gold, per acre. The soil was a rather light sandy loam, not inherently fertile, and fertilizers to the value of $3.61, gold, per acre, had been applied, leaving the earning $16.71 per acre.

Another farmer with whom we talked, pulling his crop of wheat, would follow this with millet and soy beans in alternate rows. His yield of wheat was expected to be 11 to 12 bushels per acre, his beans 21 bushels and his millet 25 bushels, which, at the local prices for grain and straw, would bring a gross earning of $35, gold, per acre.

Before reaching the end of our walk through the fields toward the next station we came across another of the many instances of the labour these people are willing to perform for only a small possible increase in crop. The field was adjacent to one of the windbreak hedges, the trees had spread their roots far afield and were threatening the crop through the consumption of moisture and plant food. To check this depletion the farmer had dug a trench 20 inches deep the length of his field, and some 20 feet from the line of trees, thereby cutting all of the surface roots to stop their draught on the soil. The trench was left open and an interesting feature observed was that nearly every cut root on the field side of the trench had thrown up one or more shoots bearing leaves, while the ends still connected with the trees showed no signs of leaf growth.

In Chihli, as elsewhere, the Chinese are skilled gardeners, using water for irrigation whenever it is advantageous. One gardener was growing a crop of early cabbage, followed by one of melons, and these with radish the same season. He was paying a rent of $6.45, gold, per acre; was applying fertilizer at a cost of nearly $8 per acre for each of the three crops, making his cash outlay $29.67 per acre. His crop of cabbage sold for $103, gold; his melons for $77, and his radish for something more than $51, making a total of $232.20 per acre, leaving him a net value of $202.53.

A second gardener, growing potatoes, obtained a yield, when sold new, of 8,000 pounds per acre; and of 16,000 pounds when the crop was permitted to mature. The new potatoes were sold so as to bring $51.60 and the mature potatoes $185.76 per acre, making the earning for the two crops the same season a total of $237.36, gold. By planting the first crop very early these gardeners secure two crops in the same season, as far north as Columbus, Ohio, and Springfield, Illinois, the first crop being harvested when the tubers are about the size of walnuts. The rental and fertilizers in this case amounted to $30.96 per acre.

Still another gardener, growing winter wheat followed by onions, and these by cabbage, both transplanted, realized from the three crops a gross earning of $176 73, gold, per acre, and incurred an expense of $31.73 per acre for fertilizer and rent, leaving him a net earning of $145 per acre.

These old people have acquired the skill and practice of storing and preserving such perishable fruits as pears and grapes so as to enable them to keep them on the markets almost continuously. Pears were very common in the latter part of June, and Consul-General Williams informed me that grapes are regularly carried into July. In talking with my interpreter as to the methods employed, I could only learn that the growers depend simply upon dry earth cellars which can be maintained at a very uniform temperature, the separate fruits being wrapped in paper. No foreigner with whom we talked knew their methods.

Vegetables are carried through the winter in such earth cellars as are seen in Fig. 73, page 144, these being covered after they are filled.

As to the price of labour in this part of China, we learned through Consul-General Williams that a master mechanic may receive 50 cents, Mexican, per day, and a journeyman 18 cents, or at a rate of 21.5 cents and 7.75 cents, gold. Farm labourers receive from $20 to $30, Mexican, or $8.60 to $12.90, gold, per year, with food, fuel and presents, which make a total of $17.20 to $21.50. This is less for the year than we pay for a month of probably less efficient labour. There is relatively little child labour in China, and this perhaps should be expected when adult labour is so abundant and so cheap.

XVI

MANCHURIA AND KOREA

The 39th parallel of latitude lies just south of Tientsin; followed westward, it crosses the toe of Italy's boot, leads past Lisbon in Portugal, near Washington and St. Louis and to the north of Sacramento on the Pacific. We were leaving a country with a mean July temperature of 80° F., and of 21° in January, but where two feet of ice may form; a country where the eighteen-year mean maximum temperature is 103·5° and the mean minimum 4·5°; where twice in this period the thermometer recorded 113° above zero, and twice 7° below, and yet near the coast and in the latitude of Washington; a country where the mean annual rainfall is 19·72 inches, of which all but 3·37 inches falls in June, July, August and September. We had taken the 5.40 a.m. Imperial North-China train, June 17th, to go as far northward as Chicago, – to Mukden in Manchuria, a distance by rail of some 400 miles, but all of the way still across the northward extension of the great Chinese coastal plain. Southward, out from the coldest quarter of the globe, where the mean January temperature is more than 40° below zero, sweep northerly winds which bring to Mukden a mean January temperature only 3° above zero, and yet there the July temperature averages as high as 77° and there is a mean annual rainfall of but 18·5 inches, coming mostly in the summer, as at Tientsin.

Although the rainfall of the northern extension of China's coastal plain is small, its efficiency is relatively high because of its most favourable distribution and the high summer temperatures. In the period of early growth, April, May and June, there are 4·18 inches; but in the period of maximum growth, July and August, the rainfall is 11·4 inches; and in the ripening period, September and October, it is 3·08 inches, while during the rest of the year only 1·06 inch falls. Thus most of the rain comes at the time when the crops require the greatest daily consumption, and it is least in mid-winter during the period of little growth.

As our train left Tientsin we travelled for a long distance through a country agriculturally poor and little tilled, with surface flat, the soil apparently saline, and the land greatly in need of drainage. Wherever there were canals the crops were best, apparently occupying more or less continuous areas along either bank. The day was hot and sultry, but labourers were busy with their large hoes, often with all garments laid aside except a short shirt and a pair of roomy trousers.

In the salt district about the village of Tangku there were huge stacks of salt and smaller piles not yet brought together, with numerous windmills, constituting most striking features in the landscape, but there was almost no agricultural or other vegetation. Beyond Pehtang there are other salt works and a canal leads westward to Tientsin, on which the salt is probably taken thither, and still other salt stacks and windmills continued visible until near Hanku, where another canal leads toward Peking. Here the coast recedes eastward from the railway, and beyond the city limits many grave mounds dot the surrounding plains where herds of sheep were grazing.

As we hurried toward the delta region of the Lwan-ho, and before reaching Tangshan, a more productive country was traversed. Thrifty trees made the landscape green, and fields of millet, kaoliang and wheat stretched for miles together along the track and back over the flat plain beyond the limit of vision. Then came fields planted with two rows of maize alternating with one row of soy beans, not more than 28 inches apart, one stalk of corn in a place every 16 to 18 inches, all carefully hoed, weedless and blanketed with an excellent earth mulch; notwithstanding which the leaves were curling in the intense heat of the sun. Tangshan is a large city, apparently of recent growth, on the railroad in a country where isolated conical hills rise 100 or 200 feet out of the flat plains. Cartloads of finely pulverized earth compost were here moving to the fields in large quantities, and laid in single piles of 500 to 800 pounds, 40 to 60 feet apart. At Kaiping the country is a little rolling and we passed through the first railway cuts, 6 to 8 feet deep, while the water in the streams was running 10 to 12 feet below the surface of the fields. On the right and beyond Kuyeh there are low hills. Here we passed enormous quantities of dry, finely-powdered earth compost, distributed on narrow unplanted areas over the fields. What crop, if indeed any, had occupied these areas this season, we could not judge. The fertilization here is even more extensive and more general than we found it in the Shantung province, and in places water was being carried in pails to the fields for use either in planting or in transplanting, to ensure the readiness of the new crops to utilize the first rainfall when it comes.

Then the bed of a nearly dry stream some 300 feet wide was crossed and beyond it a sandy plain was planted in long narrow fields between windbreak hedges. The crops were small but evidently improved by the influence of the shelter. The sand in places had drifted into the hedges to a height of 3 feet. At a number of other places along the way before Mukden was reached such protected areas were passed and oftenest on the north side of wide, now nearly dry, stream channels.

As we passed on toward Shanhaikwan we were carried over broad plains even more nearly level and unobstructed than those to be found in the corn belt of the Middle West in the United States. They too were planted with corn, kaoliang, wheat and beans, while low houses were hidden in distant scattered clusters of trees dotting the wide plain on either side, with not a fence, and nothing to suggest a road anywhere in sight. We seemed to be moving through one vast field dotted with hundreds of busy men, while here and there a great cart appeared hopelessly lost in the field, so difficult was it to trace any sign of road to guide their course.

Some early crops appeared to have been harvested from areas alternating with those under growth, and these areas were dotted with piles of the soil and manure compost, aggregating hundreds of tons, distributed over the fields, no doubt to be worked into the soil in the course of the next three or four days.

It was at Lwanchow that we met the outgoing tide of soy beans destined for Japan and Europe, pouring in from the surrounding country in gunny sacks brought on heavy carts drawn by large mules, as seen in Fig. 182. Enormous quantities had been stacked in the open along the tracks, with no shelter whatever, awaiting the arrival of trains to move them to export harbours.

The planting here, as elsewhere, is in rows, but not of one kind of grain. Most frequently two rows of maize, kaoliang or millet alternated with the soy beans, and usually not more than 28 inches apart. Such planting secures the requisite sunshine with a larger number of plants on the field; it secures a continuous general distribution of the roots of the nitrogen-fixing soy beans in the soil of the whole field every season, and permits the soil to be more continuously and more completely laid under tribute by the root systems. In places where the stand of corn or millet was too open the gaps were filled with the soy beans. Such a system of planting possibly permits a more immediate utilization of the nitrogen gathered from the soil air in the root nodules, as these die and undergo nitrification during the same season, while the crops are yet on the ground, and so far as phosphorus and potassium compounds are liberated by this decay, they too would become available for the crops.

Fig. 182. – Exportation of soy beans from Manchuria. Lwanchow, Chihli.

The end of the day's journey was at Shanhaikwan on the boundary between Chihli and Manchuria, the train stopping at 6.20 p.m. for the night. Stepping upon the veranda from our room on the second floor of a Japanese inn in the early morning, there stood before us, sullen and grey, the eastern terminus of the Great Wall, which winds 1,500 miles westward across twenty degrees of longitude, and has endured through twenty-one centuries. It is the most stupendous piece of construction ever conceived by man and executed by a nation. More than 20 feet thick at the base and more than 12 feet on the top; rising 15 to 30 feet above the ground with parapets along both faces, and towers every 200 yards rising 20 feet higher, it must have been – having in view the times and the methods of warfare then practised—when defended by their thousands, the boldest and most efficient national defence ever constructed. Nor in the economy of construction and maintenance has it ever been equalled.

Even if it be true that 20,000 masons toiled through ten years in its building, defended by 400,000 soldiers, fed by a commissariat of 20,000 more and supported by 30,000 others in the transport, quarry and potters' service, she would then have been using less than eight-tenths per cent of her population, reckoned as 60,000,000 at that time; while, according to Edmond Théry's estimate, the officers and soldiers of Europe to-day, in time of peace, constitute 1 per cent of a population of 400,000,000 of people, and these, at only one dollar each per day for food, clothing and loss of producing power, would cost her nations, in ten years, more than $14,000 million. China, with her present habits and customs, would more easily have maintained her army of 470,000 men on 30 cents each per day, or for a total ten-year cost of but $520,000,000. The French cabinet in 1900 approved a naval programme involving an expenditure of $600,000,000 during the next ten years, a tax of more than $15 for every man, woman and child in the Republic.

Leaving Shanhaikwan at 5.20 in the morning and reaching Mukden at 6.30 in the evening, we rode the entire day through Manchurian fields. Manchuria has an area of 363,700 square miles, equal to that of both Dakotas, Minnesota, Nebraska and Iowa combined. It has roughly the outline of a huge boot, and could one slide it eastward until Port Arthur was at Washington, Shanhaikwan would fall well toward Pittsburg, both at the tip of the broad toe to the boot. The foot would lie across Pennsylvania, New York, New Jersey and all of New England, extending beyond New Brunswick with the heel in the Gulf of St. Lawrence. Harbin, at the instep of the boot, would lie 50 miles east of Montreal and the expanding leg would reach north-westward nearly to James Bay, entirely to the north of the Ottawa river and the Canadian Pacific, spanning 1,000 miles of latitude and 900 miles of longitude.

The Liao plain, 30 miles wide, and the central Sungari plain are the largest in Manchuria, forming together a long narrow valley floor between two parallel mountain systems and extending north-easterly from the Liao gulf, between Port Arthur and Shanhaikwan, up the Liao river and down the Sungari to the Amur, a distance of 800 or more miles. These plains have a fertile, deep soil, and it is on them and other lesser river bottoms that Manchurian agriculture is developed, supporting eight or nine million people on a cultivated acreage possibly not greater than 25,000 square miles.

Manchuria has great forest and grazing possibilities awaiting future development, as well as much mineral wealth. The population of Tsitsihar, in the latitude of middle North Dakota, swells from 30,000 to 70,000 during September and October, when the Mongols bring their cattle in to market. In the middle province, at the head of steam navigation on the Sungari, because of the abundance and cheapness of lumber, Kirin has become a shipbuilding centre for Chinese junks. The Sungari – Milky – river is a large stream carrying more water at flood season than the Amur above its mouth, the latter being navigable 450 miles for steamers drawing 12 feet of water, and 1,500 miles for those drawing 4 feet, so that during the summer season the middle and northern provinces have natural inland waterways, but the outlet to the sea is far to the north and closed by ice six months of the year.

Not far beyond the Great Wall of China, fast falling into ruin, partly through the appropriation of its material for building purposes now that it has outlived its usefulness, another broad, nearly dry stream-bed was crossed. There, in full bloom, was what appeared to be the wild white rose seen earlier, farther south, west of Suchow, having a remarkable profusion of small white bloom in clusters resembling the Rambler rose. One of these bushes growing wild there on the bank of the canal had overspread a clump of trees, one of which was 30 feet in height, enveloping it in a mantle of bloom, as seen in the upper section of Fig. 183. The lower section of the illustration is a closer view showing the clusters. The stem of this rose, 3 feet above the ground, measured 14·5 inches in circumference. If it would thrive in western countries nothing could be better for parks and pleasure drives. Later on our journey we saw it many times in bloom along the railway between Mukden and Antung, but nowhere attaining so large a growth. The blossoms are scant three-fourths of an inch in diameter, usually in compact clusters of three to eleven, sometimes in twos and occasionally standing singly. The leaves are five-foliate, sometimes trifoliate; leaflets broadly lanceolate, accumi-

Fig. 183. – Wild white rose in bloom west of Suchow, June 2nd, and in southern Manchuria, June 18th. Lower section, close view of same, showing clusters.

nate and finely serrate; thorns minute, recurrent and few, only on the smaller branches.

In a field beyond, a small donkey was drawing a stone roller 3 feet long and 1 foot in diameter, firming the crests of narrow, sharp, recently formed ridges, two at a time. Millet, maize and kaoliang were here the chief crops. Another nearly dry stream was crossed, where the fields became more rolling and much cut by deep gulleys, the first instances we had seen in China except on the steep hill-sides about Tsingtao. Not all of the lands here were cultivated, and on the untilled areas herds of fifty to a hundred goats, pigs, cattle, horses and donkeys were grazing.

Fields in Manchuria are larger than in China and some rows were a full quarter of a mile long, so that cultivation was being done with donkeys and cattle, and large numbers of men were working in gangs of four, seven, ten, twenty, and in one field as high as fifty, hoeing millet. Such a crew as the largest mentioned could probably be hired at 10 cents each, gold, per day, and were probably men from the thickly settled portions of Shantung who had left in the spring, expecting to return in September or October. Both labourers and working animals were taking dinner in the fields, and earlier in the day we had seen several instances where hay and feed were being taken to the field on a wooden sled, with the plough and other tools. At noon this was serving as manger for the cattle, mules or donkeys.

In fields where the close, deep furrowing and ridging was being done the team often consisted of a heavy ox and two small donkeys driven abreast, the three walking in adjacent rows, the plough following the ox, or a heavy mule instead.

The rainy season had not begun and in many fields there was planting and transplanting where water was used in separate hills, sometimes brought in pails from a near-by stream, and in other cases on carts provided with tanks. Holes were made along the crests of the ridges with the blade of a narrow hoe and a little water poured in each hill, from a dipper, before planting or setting. These were additional instances of the farmers' willingness to incur additional labour to save time for the maturing of the crop by assisting germination in a soil too dry to make it certain, until the rains came.

It appears probable that the strong ridging and the close level rows so largely adopted here must have marked advantages in utilizing the rainfall, especially that portion of it which comes early, and that which comes later also if it should come in heavy showers. With steep narrow ridging, heavy rains would be shed at once to the bottom of the deep furrows without over-saturating the ridges, while the wet soil in the bottom of the furrows would favour deep percolation with lateral capillary flow under the ridges from the furrows, carrying both moisture and soluble plant food where they will be most completely and quickly available. When the rain comes in heavy showers each furrow may serve as a long reservoir, which will prevent washing and at the same time permit quick penetration. The ridges never becoming flooded or puddled, permit the soil air to escape readily as the water from the furrows sinks. This it cannot easily do in flat fields when the rains fall rapidly and fill all the soil pores, because when this happens the soil pores are closed to the escape of air from below, which must take place before the water can enter.

When rows are only 24 to 28 inches apart, ridging is not sufficiently wasteful of soil moisture – because of the greater evaporation due to increased surface – to compensate for the other advantages gained, and hence the practice described in the preceding paragraph, for these conditions, appears sound.

The application of finely pulverized earth compost to fields about to be planted, and in some cases where the fields were already planted, continued general after leaving Shanhaikwan, as it had been before. Compost stacks were common in yards wherever buildings were close enough to the track to be seen. Much of the way about one-third of the fields were yet to be, or had just been, planted, and in a great majority of these compost fertilizer had been laid down for use on them, or was being taken to them in large heavy carts drawn sometimes by three mules. Between Sarhongon and Ningyuenchow fourteen fields thus fertilized were counted in less than half a mile; ten others in the next mile; eleven in the mile and a quarter following. In the next two miles 100 fields were counted, and just before reaching the station we counted during five minutes, with watch in hand, ninety-five fields to be planted, upon which this fertilizer had been brought. In some cases the compost was being spread in furrows between the rows of a last year's crop, evidently to be turned under, thus reversing the position of the ridges.

After passing Lienshan, where the railway runs near the sea, a sail was visible and many stacks of salt piled about the evaporation fields were associated with the revolving sail windmills already described. Here, too, large numbers of cattle, horses, mules and donkeys were grazing on the untilled lowlands, beyond which we traversed a section where all fields were planted, where no fertilizer was piled in the field, but where many groups of men were busy hoeing, sometimes twenty in a gang.

Chinese soldiers with bayoneted guns stood guard at every railway station between Shanhaikwan and Mukden, and from Chinchowfu our coach was occupied by a Chinese official with guests and military attendants, including armed soldiers. The official and his guests were an attractive group of men with pleasant faces and winning manners, clad in many garments of richly-figured silk of bright, attractive, but unobtrusive, colours, who talked, seriously or in mirth, almost incessantly. They took the train about one o'clock and lunch was immediately served in Chinese style, but the last course was not brought until nearly four o'clock. At every station soldiers stood in line in the attitude of salute until the official car had passed.

Just before reaching Chinchowfu we saw the first planted fields littered with stubble of the previous crop, and in many instances such stubble was being gathered and removed to the villages, large stacks having been piled in the yards to be used either as fuel or in the production of compost. As the train approached Taling-ho groups of men were hoeing in millet fields, thirty in one group on one side and fifty in another group on the other. Many small herds of cattle, horses, donkeys and flocks of goats and sheep were feeding along stream courses and on the unplanted fields. Beyond the station, after crossing the river, still another sand dune tract was passed, planted with willows, millet occupying the level areas between the dunes. Not far beyond, wide untilled flats on which many herds were grazing were crossed. As we neared Koupantze, where a branch of the railway traverses the Liao plain to the port of Newchwang, they were dotted with grave mounds. It was in this region that there came the first suggestion of resemblance to our marshland meadows; and very soon there were seen approaching from the distance loads so green that except for the large size one would have judged them to be fresh grass. They were loads of cured hay in the brightest green, the result, no doubt, of curing under their dry weather conditions.

At Ta Hu Shan large quantities of grain in sacks were piled

along the tracks and in the freight yards, but under matting shelters. Near here, too, large three-mule loads of dry earth compost were going to the fields and men were busy pulverizing and mixing it on the threshing-floors preparatory for use. Nearly all crops growing were one or another of the millets, but considerable areas were yet unplanted, and on these cattle, horses, mules and donkeys were feeding,

When the train reached Sinminfu, where the railway turns abruptly eastward to cross the Liao-ho to reach Mukden, we saw the first extensive massing of the huge bean cakes for export, together with enormous quantities of soy beans in sacks piled along the railway, or loaded on cars made up in trains ready to move. Leaving this station we passed among fields of grain looking decidedly yellow, the first indication we had seen in China of crops nitrogen-hungry and of soils markedly deficient in available nitrogen. Beyond the next station the fields were spotted and uneven as well as yellow, recalling conditions so commonly seen at home and which had hitherto been conspicuously absent here. We then crossed the Liao-ho with its broad channel of shifting sands. The river carried the largest volume of water we had yet seen, although the stream was very low and still, characteristic of the close of the dry season of semi-arid climates. We soon reached another station where the freight yards and all of the space along the tracks were piled high with bean cakes.

Since the Japanese-Russian war the shipments of soy beans and of bean cake from Manchuria have increased enormously. Up to this time there had been exports to the southern provinces of China, where the bean cakes were used as fertilizers for the rice fields, but the new extensive markets have so raised the price that in several instances we were informed they could not now afford to use bean cake for this purpose. From Newchwang alone, in 1905, between January 1st and March 31st, there went abroad 2,286,000 pounds of bean and bean cake, and in 1906 the amount had increased to 4,883,000 pounds. But a report published in the Tientsin papers as official stated that the value of the export of bean cake and soy beans from Dalny for the months ending March 31st had been, in 1909, only $1,635,000, gold, compared with $3,065,000 in the corresponding period of 1908, and of $5,120,000 in 1907, showing a marked decrease.

Edward C. Parker, writing from Mukden for the *Review of Reviews*, stated: 'The bean cake shipments from Newchwang, Dalny and Antung in 1908 amounted to 515,198 tons; beans, 239,298 tons; bean oil, 1,930 tons; having a total value of $15,016,649 (U.S. gold).'

According to the composition of soy beans as indicated in Hopkins' table of analyses, these shipments of beans and bean cake would remove an aggregate of 6,171 tons of phosphorus, 10,097 tons of potassium, and 47,812 tons of nitrogen from Manchurian soils as the result of export for that year. Could such a rate have been maintained during two thousand years, there would have been sold from these soils 20,194,000 tons of potassium, 12,342,000 tons of phosphorus, and 95,624,000 tons of nitrogen; and the phosphorus, were it thus exported, would have exceeded more than threefold all thus far produced in the United States; it would have exceeded the world's output in 1906 more than eighteen times, even assuming that all phosphate rock mined was 75 per cent pure.

The choice of the millets and the sorghums as the staple bread crops of northern China and Manchuria has been quite as remarkable as the selection of rice for the more southern latitudes, and the two together have played a most important part in determining the high maintenance efficiency of these people. In nutritive value these grains rank well with wheat; the stems of the larger varieties are extensively used for both fuel and building material, while the smaller forms make excellent forage and have been used directly for maintaining the organic content of the soil. Their rapid development and their high endurance of drought adapt them admirably to the climate of north China and Manchuria, where the rains begin only after late June and where weather too cold for growth comes earlier in the autumn. The quick maturity of these crops also permits them to be used to great advantage even in the south, where the system of multiple cropping is so generally adopted, while their great resistance to drought – enabling them to remain at a standstill for a long time when the soil is too dry for growth and yet to push ahead rapidly when favourable rains come – permits them to be used on the higher lands where water is not available for irrigation.

In the Shantung province the large millet, sorghum or kaoliang, yields as high as 2,000 to 3,000 pounds of seed per acre, and 5,600 to 6,000 pounds of air-dry stems, equal in weight to 1·6 to 1·7 cords of dry oak wood. In the region of Mukden, Manchuria, its average yield of seed is placed at 35 bushels of 60 pounds weight per acre, and with this comes 1½ tons of fuel or of building material. Hosie states that the kaoliang is the staple food of the population of Manchuria and the principal grain food of the work animals. The grain is first washed in cold water and then poured into a kettle with four times its volume of boiling water and cooked for an hour, without salt, as with rice. It is eaten with chopsticks with boiled or salted vegetables. He states that an ordinary servant requires about 2 pounds of this grain per day, and that a workman at heavy labour will take double the amount. A Chinese friend of his, keeping five servants, supplied them with 240 pounds of millet per month, together with 16 pounds of native flour for two days, and meat for two days, the amount not being stated. Two of the small millets (*Setaria Italica* and *Panicum milliaceum*), wheat, maize and buckwheat are other grains which are used as food, but chiefly to give variety and change of diet.

Large quantities of matting and wrappings are also made from the leaves of the large millet, which serve many purposes, corresponding with the rice mattings and bags of Japan and southern China.

The small millets, in Shantung, yield as high as 2,700 pounds of seed and 4,800 pounds of straw per acre. In Japan, in the year 1906, there were grown 737,719 acres of foxtail, barnyard and proso millet, yielding 17,084,000 bushels of seed, or an average of 23 bushels per acre. In addition to the millets, Japan grew, the same year, 5,964,300 bushels of buckwheat on 394,523 acres, or an average of 15 bushels per acre. The next engraving, Fig. 184, shows a crop of millet already 6 inches high planted between rows of windsor beans which had matured about the middle of June. The leaves had dropped, the beans had been picked from the stems, and a little later, when the roots had had time to decay, the bean stems would be pulled and tied in bundles for use as fuel or for fertilizer.

We had reached Mukden thoroughly tired after a long day of continuous close observation and writing. The Astor House, where we were to stop, was three miles from the station and the only conveyance to meet the train was a four-seated springless, open, semi-baggage carry all, and it was a full hour lumbering its way to our hotel. But here, as everywhere in the Orient, the foreigner meets scenes and phases of life competent to divert his attention from almost any discomfort. Nothing could be more striking than the peculiar mode the Manchu ladies have of dressing their hair, many instances of which were passed on the streets during this early evening ride. It was fearfully and wonderfully done,

FIG. 184. – Field of millet planted between rows of windsor beans. Chiba, Japan.

laid in the smoothest, glossiest black, with nearly the lateral spread of the tail of a turkey-cock and much of the backward curve of that of the rooster; far less attractive than the plainer, refined, modest, yet highly artistic style adopted by either Chinese or Japanese ladies.

The journey from Mukden to Antung required two days, the train stopping for the night at Tsaohokow. Our route lay most of the way through mountainous or steep hilly country and our train was made up of diminutive coaches. They were drawn by a tiny engine along a 3-foot 2-inch narrow-gauge track of light rails laid by the Japanese during the war with Russia, for the purpose of moving their armies and supplies to the hotly-contested fields in the Liao and Sungari plains. Many of the grades were steep, the curves sharp, and in several places it was necessary to divide the short train to enable the engines to negotiate them.

Southward over the Liao plain the crops were almost exclusively millet and soy beans, with a little barley, wheat, and a few oats. Between Mukden and the first station across the Hun river we had passed twenty-four good-sized fields of soy beans on one side of the river and twenty-two on the other, and before reaching the hilly country, after travelling a distance of possibly 15 miles, we had passed 309 other and similar fields close along the track. In this distance also we had passed two of the monuments erected by the Japanese, marking sites of their memorable battles. These fields were everywhere flat, lying from 16 to 20 feet above the beds of the nearly dry streams, and the cultivation was mostly done with horses or cattle.

After leaving the plains country the railway traversed a narrow winding valley less than a mile wide, with gradient so steep that our train was divided. Fully 60 per cent of the hill-slopes were cultivated nearly to the summit, although rising apparently more than 1 in 3 to 5 feet. The uncultivated slopes were closely wooded with young trees, few more than 20 to 30 feet high, but in blocks evidently of different ages. Beyond the pass many of the cultivated slopes have walled terraces. We crossed a large stream where railway ties were being rafted down the river. Just beyond this river the train was again divided to ascend a gradient of one in thirty, reaching the summit by five times switching back, and matched on the other side of the pass by a down grade of one in forty.

At many of the farm-houses in the narrow valleys along the way large rectangular, flat-topped compost piles were passed, 30 to 40 inches high and 20, 30, 40 and even in one case as much as 60 feet square on the ground. More and more it became evident that these mountain and hill lands were originally heavily wooded and that the new growth springs up quickly, developing rapidly. It was clear also that the custom of cutting over these wooded areas at frequent intervals is very old, not always in the same stage of growth but usually when the trees are quite small. Considerable quantities of cordwood were piled at the stations along the railway and were being loaded on the cars. This was always either round wood or sticks split but once; and much charcoal, made mostly from round wood or sticks split but once, was being shipped in sacks shaped like those used for rice, seen in Fig. 161. Some strips of the forest growth had been allowed to stand undisturbed apparently for twenty or more years, but most areas have been cut at more frequent intervals, often once in three to five, or perhaps ten, years.

At several places on the rapid streams crossed, prototypes of the modern turbine water-wheel were installed for grinding beans or grain. As with native machinery everywhere in China, these wheels were reduced to the lowest terms and the principle set to work almost unclothed. The turbines were of the downward discharge type, much resembling our modern windmills, 10 to 16 feet in diameter, set horizontally on a vertical axis rising through the floor of the mill, with the vanes surrounded by a rim, the water dropping through the wheel, reacting when reflected from the obliquely set vanes. American engineers and mechanics would pronounce these crude, primitive and inefficient. A truer view would regard them as examples of a masterful grasp of principle by some man who long ago saw the unused energy of the stream and succeeded thus in turning it to account.

Both days of our journey had been bright and very warm, and, although we took the train early in the morning at Mukden, a young Japanese, anticipating the heat, entered the train clad only in his kimono and sandals, carrying a suit-case and another bundle. He rode all day, the most comfortably, if immodestly, clad man on the train. The next morning he took his seat in front of us clad in the same garb, but before the train reached Antung he took down his suit-case and then and there attired himself in a good foreign suit, folding his kimono and packing it away with his sandals.

From Antung we crossed the Yalu on the ferry to New Wiju at 6.30 a.m., June 22nd. We found ourselves in quite a different country and among a very different people, although all of the railway officials, employees, police and guards were Japanese, as they had been from Mukden. At Antung and New Wiju the Yalu is a broad slow stream resembling an arm of the sea more than a river, reminding one of the St. Johns at Jacksonville, Florida.

June 22nd proved to be one of the national festival days in Korea, called 'Swing day,' and throughout our entire ride to Seoul the fields were nearly all deserted and throngs of people, arrayed in gala dress, appeared all along the line of the railway, sometimes congregating in bodies of two to three thousand or more. Many swings had been hung and were being enjoyed by the young people. Boys and men were bathing in all sorts of 'swimming holes' and places. So, too, there were many large open-air gatherings being addressed by public speakers.

Nearly every one was dressed in white outer garments made from some fabric which, although not mosquito netting, was nearly as open. It was possessed of a remarkable stiffness, which seemed to take and retain every dent with astonishing effect and was sufficiently transparent to reveal a third undergarment. The women wore full outstanding skirts, and the trousers which went with these were proportionately full but tied close about the ankles. The garments seemed to be possessed of a powerful repulsion which held them quite apart and away from the person, no doubt contributing much to comfort. It was windy, but hot, sultry and sticky, and it made one feel cool to see these open garments surging in the wind.

The Korean men, like the Chinese, wear the hair long but not braided in a queue. No part of the head is shaved, but the hair is wound in a tight coil on the top of the head, secured by a pin which, in the case of the Korean who rode in our coach from Mukden to Antung, was a modern, substantial tenpenny wire nail. The tall, narrow, conical crowns of the open hats, woven from thin bamboo splints, are evidently designed to accommodate this style of hair-dressing as well as to be cool.

Here, as in China and Manchuria, nearly all crops are planted in rows, including the cereals, such as wheat, rye, barley and oats. We traversed first a flat marshy country with sandy soil and water not more than 4 feet below the surface where, on the lowest areas a close ally of our wild flower-de-luce was in bloom. Wheat was coming into head, but corn and millet were smaller than in Manchuria. We had left New Wiju at 7.30 in the morning and at 8.15 we passed from the low land into a hill country with narrow valleys.

Scattering young pine, seldom more than 10 to 25 feet high, occupied the slopes and, as we came nearer, the hills were seen to be clothed with many small oak, the sprouts clearly not more than one or two years old. Roofs of dwellings in the country were usually thatched with straw, laid after the manner of shingles, as may be seen in Fig. 185, where the hills beyond show the low tree growth referred to, but unusually dense. Bundles of pine boughs,

Fig. 185. – Group of Korean farm-houses, with thatched roofs and earthen walls, standing at the foot of wooded hills.

stacked and sheltered from the weather, were common along the way and evidently used for fuel.

At 8.25 we passed through the first of many tunnels along the route, the longest requiring thirty seconds for the passing of the train. The valley beyond was occupied by fields of wheat where beans were planted between the rows. Thus far none of the fields had been as thoroughly tilled and well cared for as those seen in China, nor were the crops as good. Farther along we passed hills where the pines were all of two ages, one set about 30 feet high and the others 12 to 15 feet or less, among which were numerous oak sprouts. Quite possibly these are used as food for the wild silkworms. In some places appearances indicate that the oak and other deciduous growth, with the grass, may be cut annually and only the pines allowed to stand for longer periods. As we proceeded southward and passed Kosui the young oak sprouts were seen to cover the hills, often stretching over the slopes much like a regular crop, standing at a height of 2 to 4 feet. Fresh bundles

Fig. 186. – General view across valley, showing Korean rice fields being transplanted, and in the foreground fertilized with green herbage from the hill lands.

of these sprouts were seen at houses along the foot of the slopes, again suggesting that the leaves may be for the tussur silkworms although the time appears late for the first moulting. After we had left Seoul, entering the broader valleys where rice was more extensively grown, the use of oak boughs and green grass brought down from the hill lands for green manure was very extensive.

After the winter and early spring crops have been harvested the narrow ridges on which they are grown are turned into furrows by means of a simple plough drawn by a heavy bullock, different from the cattle in China but closely similar to those in Japan. The fields are then flooded until they have the appearance seen in Fig. 10. Over these flooded ridges the green grass and oak boughs are spread, when the fields are again ploughed and the material worked into the wet soil. If this working is not completely successful, men enter the fields and tramp the surface until every twig and blade is submerged. The middle section in the same illustration has been fitted and transplanted; in front of it and on the left are two other fields once ploughed but not fertilized; these far to the right have had the green manure applied and the ground ploughed a second time but not finished, and in the imme-

Fig. 187. – Rice field covered with oak leaves and grass brought down from the hills, one half of which has been tramped beneath the surface by the labourer at work.

diate foreground the grass and boughs have been scattered but the second ploughing is not yet done.

We passed men and bullocks coming from the hill lands loaded with this green herbage, and as we proceeded towards Fusan more and more of the hill area was being made to contribute materials for green manure for the cultivated fields. The foreground of Fig. 186 had been thus treated and so had the field in Fig. 187, where the man was engaged in tramping the dressing beneath the surface. In very many cases this material was laid along the margin of the paddy fields; in other cases it had been taken upon the fields as soon as the grain was cut and was lying in piles among the bundles; while in still other cases the material for green manure had been carried between the rows while the grain was still standing, but nearly ready to harvest. In some fields a full third of a bushel of the green stuff had been laid down at intervals of 3 feet over the whole area. In others piles of ashes alternated with those of herbage, and again manure and ashes mixed had been distributed in alternate piles with the green manure; and in others again we saw untreated straw distributed through the fields awaiting application. At Shindo this straw had the appearance of having been dipped in or smeared with some mixture, apparently of mud and ashes or possibly of some compost which had been worked into a thin paste with water.

After passing Keizan, mountain herbage had been brought down from the hills in large bales on cleverly constructed racks saddled to the backs of bullocks, and in one field we saw a man who had just come to his little field with an enormous load borne upon his easel-like packing appliance. Thus we find the Koreans also adopting the rice crop, which yields heavily under conditions of abundant water; we find them supplementing a heavy summer rainfall with water from their hills, and bringing to their fields besides both green herbage for humus and organic matter, and ashes derived from the fuel which again comes from the hills. In these ways they make good the losses unavoidable through intense cropping.

The amount of forest growth in Korea, as we saw it, in proximity to the cultivated valleys, is nowhere large and is fairly represented in Figs. 185, 188 and 189. There were clear evidences of periodic cutting and considerable amounts of cordwood split from timber a foot through were being brought to the stations on the backs of animals. In some places there was evident very serious soil erosion, as may be seen in Fig. 189. One such region was passed just before reaching Kinusan, but generally the hills are well rounded and covered with a low growth of shrubs and herbaceous plants.

Southernmost Korea has the latitude of the northern boundary of South Carolina, Georgia, Alabama and Mississippi, while the north-east corner attains that of Madison, Wisconsin, and the northern boundary of Nebraska, the country thus spanning some nine degrees or 600 miles of latitude. It has an area of some 82,000 square miles, about equalling the State of Minnesota, but much of its surface is occupied by steep hill and mountain land. The

Fig. 188. – Rice fields at head of mountain valley, with scattering pines in the hill lands beyond.

Fig. 189. – Looking across fields of wheat at an eroding hillside over which forest growth is being allowed to spread.

rainy season had not yet set in, June 23rd. Wheat and the small grains were practically all harvested southward of Seoul and the people were everywhere busy with their flails threshing in the open, about the dwellings or in the fields, four flails often beating together on the same lot of grain. As we journeyed southward the valleys and the fields became wider and more extensive, and the crops, as well as the cultural methods, were clearly better.

Fig. 190. – Korean swinging scoop for irrigation where the water is raised 3 or 4 feet.

Neither the foot-power, animal-power, nor the wooden chain pump of the Chinese were observed in Korea for lifting water, but we saw many instances of the long-handled, spoonlike swinging scoop hung over the water by a cord from tall tripods, after the manner seen in Fig. 190, each operated by one man and apparently with high efficiency for low lifts. Two instances also were observed of the form of lift seen in Fig. 155, where the man walks the circumference of the wheel, so commonly observed in Japan. Much hemp was being grown in southern Korea but everywhere on very small isolated areas which flecked the landscape with the deepest green, each little field probably representing the crop of a single family.

It was 6.30 p.m. when our train reached Fusan after a hot and dusty ride. The service had been good and fairly comfortable but the ice-water tanks of American trains were absent, their place being supplied by cooled bottled waters of various brands, including soda-water, sold by Japanese boys at nearly every important station. Close connection was made by trains with steamers to and from Japan and we went immediately on board the *Iki Maru* which was to weigh anchor for Moji and Shimonoseki at 8 p.m. Although small, the steamer was well equipped, providing the best of service. We were fortunate in having a smooth passage, anchoring at 6.30 the next morning and making close connection with the train for Nagasaki, landing at the wharf with the aid of a steam launch.

Our ride by train through the island of Kyushu carried us through scenes not widely different from those we had just left. The journey was continuously among fields of rice, with Korean features strongly marked but usually under better and more intensified culture, and the season, too, was a little more advanced. Here the ploughing was done mostly with horses instead of the heavy bullocks so exclusively employed in Korea. Coming from China into Korea, and from there into Japan, it appeared very clear that in agricultural methods and appliances the Koreans and Japanese are mutually more similar than the Chinese and Koreans, and the more we came to see of the Japanese methods the more strongly the impression became fixed that either the Japanese had derived their methods from the Koreans or the Koreans had taken theirs from Japan more than from China.

It was on this ride from Moji to Nagasaki that we were introduced to the attractive and very satisfactory manner of serving lunches to travellers on the trains in Japan. At important stations hot tea is brought to the car windows in small glazed earthenware teapots provided with cover and bail, and accompanied with a teacup of the same ware. The set and contents could be purchased for 5 sen, 2½ cents U.S. currency. All tea is served without milk or sugar. The lunches were substantial and put together in a neat sanitary manner in a three-compartment wooden box, carefully made from clear lumber joined with wooden pegs and perfect joints. Packed in the cover we found a paper napkin, toothpicks and a pair of chopsticks. In the second compartment there were thin slices of meat, chicken and fish, together with bamboo sprouts, pickles, cakes and small bits of salted vegetables, while the lower and chief compartment was filled with rice cooked stiff and without salt, as is the custom in the three countries. The box was about 6 inches long, 4 inches deep and 3½ inches wide. These lunches are handed to travellers neatly wrapped in spotless thin white paper daintily tied with a bit of colour, all in exchange for 25 sen – 12·5 cents. Thus for 15 cents the traveller is handed, through the car window, in a respectful manner, a square meal which he may eat at his leisure.

XVII

RETURN TO JAPAN

We had returned to Japan in the midst of the first rainy season, and all the day through, June 25th, and two nights, a gentle rain fell at Nagasaki. Across the narrow street from Hotel Japan were two of its guest houses, standing near the front of a wall-faced terrace rising 28 feet above the street and facing the beautiful harbour. They were accessible only by winding stone steps between retaining walls overhung with a wealth of shrubbery green and looking almost liquid in the drip of the rain. Over such another series of steps, but longer and more winding, we found our way to the American Consulate where in beautifully secluded quarters Consul-General Scidmore escaped many of the annoyances of settling imagined petty grievances arising between American tourists and the ricksha boys.

Through the kind offices of the Imperial University of Sapporo and of the National Department of Agriculture and Commerce, Professor Tokito met us at Nagasaki, to act as escort through most of the journey in Japan. Our first visit was to the prefectural Agricultural Experiment Station at Nagasaki. There are forty others in the four main islands, one to an average area of 4,280 square miles, and to each 1,200,000 people.

The island of Kyushu, whose latitude is that of middle Mississippi and north Louisiana, has two rice harvests, and gardeners at Nagasaki grow three crops each year. The gardener and his family work about 5 tan, or a little less than $1\frac{1}{4}$ acres, realizing an annual return of some $250 per acre. To maintain these earnings fertilizers are applied rated worth $60 per acre, divided between the three crops, the materials used being largely the wastes of the city, animal manure, mud from the drains, fuel ashes and sod, all composted together. If this expenditure for fertilizers appears high it must be remembered that nearly the whole product is sold and that there are three crops each year. Such intense culture requires a heavy return if large yields are to be maintained. Good agricultural lands were here valued at 300 yen per tan, approximately $600 per acre.

When returning toward Moji to visit the Agricultural Experiment Station of Fukuoka prefecture, the rice along the first portion of the route was standing about 8 inches above the water. Large lotus ponds along the way occupied areas not readily drained, and the fringing fields between the rice fields and the untilled hill lands were bearing squash, maize, beans and Irish potatoes. Many small areas had been set to sweet potatoes on close narrow ridges, the tops of which were thinly strewn with green grass, or sometimes with straw or other litter, for shade and to prevent the soil from washing and baking in the hot sun after rains. At Kitsu we passed near Government salt works, for the manufacture of salt by the evaporation of sea water, this industry in Japan, as in China, being a Government monopoly.

FIG. 191. – Working green herbage into a flooded paddy field for green manure, preparatory for the following crop of rice.

Many bundles of grass and other green herbage were collected along the way, gathered for use in the rice fields. In other cases the green manure had already been spread over the flooded paddy fields and was being worked beneath the surface, as seen in Fig. 191. At this time the hill lands were clothed in the richest, deepest green, but the tree growth was nowhere large except immediately about temples, and was usually in distinct small areas with sharp boundaries occasioned by differences in age. Some tracts had been very recently cut; others were in their second, third or fourth years; while others still carried a growth of perhaps seven to ten years. At one village many bundles of the brush fuel had been gathered from an adjacent area, recently cleared.

A few fields were still bearing their crop of soy beans planted in February between rows of grain, and the green herbage was being worked into the flooded soil, for the crop of rice. Much compost, brought to the fields, was stacked with layers of straw between, laid straight, the alternate courses at right angles, holding the piles in rectangular form with vertical sides, some of which were 4 to 6 feet high and the layers of compost about 6 inches thick.

Just before reaching Tanjiro a region is passed where orchards of the candleberry tree occupy high levelled areas between rice fields, after the manner described for the mulberry orchards in Chekiang, China. These trees, when seen from a distance, have the appearance of our apple orchards.

At the Fukuoka Experiment Station we learned that the usual depth of ploughing for the rice fields is 3½ to 4½ inches, but that deeper ploughing gives somewhat larger yields. As an average of five years' trials, a depth of 7 to 8 inches increased the yield from 7 to 10 per cent over that of the usual depth. In this prefecture grass from the bordering hill lands is applied to the rice fields at rates ranging from 3,300 to 16,520 pounds green weight per acre, and, according to analyses given, these amounts would carry to the fields from 18 to 90 pounds of nitrogen, 12·4 to 63·2 pounds of potassium, and 2·1 to 10·6 pounds of phosphorus per acre.

Where bean cake is used as a fertilizer the applications may be at the rate of 496 pounds per acre, carrying 33·7 pounds of nitrogen, nearly 5 pounds of phosphorus and 7·4 pounds of potassium. The earth composts are chiefly applied to the dry land fields and then only after they are well rotted. The fermentation is carried on through at least sixty days, during which the material is turned three times for aeration. When used on the rice fields where water is abundant the composts are applied in a less fermented condition.

The best yields of rice in this prefecture are some 80 bushels per acre, and crops of barley may even exceed this, the two crops being grown the same year, the rice following the barley. In most parts of Japan the grain food of the labouring people is about 70 per cent naked barley mixed with 30 per cent of rice, both cooked and used in the same manner. The barley has a lower market value and its use permits a larger share of the rice to be sold as a money crop.

The soils are fertilized for each crop every year and the prescription for barley and rice recommended by the Experiment Station, for growers in this prefecture, is indicated by the following table:

FERTILIZATION FOR NAKED BARLEY

Fertilizers.		Pounds per acre N	P	K
Manure compost	6,613	33·0	7·4	33·8
Rape seed cake	330	16·7	2·8	3·5
Night soil	4,630	26·4	2·6	10·2
Superphosphate	132	..	9·9	..
Sum	11,705	76·1	22·7	47·5

FERTILIZATION FOR PADDY RICE

Manure compost	5,291	26·4	5·9	27·1
Green manure, soy beans	3,306	19·2	1·1	19·6
Soy bean cake	397	27·8	1·7	6·4
Superphosphate	198	..	12·8	..
Sum	9,192	73·4	21·5	53·1
Total for year	20,897	149·5	44·2	100·6

Where these recommendations are followed there is an annual application of fertilizer material which aggregates some 10 tons per acre, carrying about 150 pounds of nitrogen, 44 pounds of phosphorus and 100 pounds of potassium. The crop yields which have been associated with these applications on the Station fields are about 49 bushels of barley and 50 bushels of rice per acre.

The general rotation recommended for this portion of Japan covers five years and consists of a crop of wheat or naked barley the first two years with rice as the summer crop; in the third year *genge*, 'pink clover' (*Astragalus sinicus*) or some other legume for green manure is the winter crop, rice following in the summer; the fourth year rape is the winter crop, from which the seed is saved and the ash of the stems returned to the soil, or rarely the stems themselves may be turned under; on the fifth and last year of the rotation the broad kidney or windsor bean is the winter crop, preceding the summer crop of rice. This rotation is not general yet in the practice of the farmers of the section, who choose rape or barley and in February plant windsor or soy beans between the rows for green manure.

It was evident from our observations that the use of composts in fertilizing was very much more general and extensive in China than it was in either Korea or Japan, but, to encourage the production and use of compost fertilizers, this and other prefectures have provided subsidies which permit the payment of $2.50 annually to those farmers who prepare and use on their land a compost heap covering 20 to 40 square yards, in accordance with specified directions.

The agricultural college at Fukuoka was not in session the day of our visit, it being a holiday usually following the close of the last transplanting season. Fig. 192 furnishes a view of the station grounds and buildings with something of the beautiful landscape setting. There is nowhere in Japan the lavish expenditure of money on elaborate and imposing architecture which characterizes American colleges and stations, but in equipment for research work, both as to professional staff and appliances, they compare favourably with similar institutions in America. The dormitory system was in vogue in the college, providing room and board at 8 yen per month, or $4 of U.S. currency. Eight students were assigned to one commodious room, each provided with a study table, but beds were mattresses spread upon the matting floor at night and compactly stored on closet shelves during the day.

The Japanese plough, which is very similar to the Korean type, may be seen in Fig. 193, the one on the right costing 2·5 yen and the other 2 yen. With the aid of the single handle and the sliding rod held in the right hand, the course of the plough is directed and the plough tilted in either direction, throwing the soil to the right or the left.

The nursery beds for rice breeding experiments and variety tests by this station are shown in Fig. 194. Although these plots were flooded, the marginal plants, adjacent to the free water paths, were materially larger than those within and had a much deeper green colour, showing better feeding, but what seemed

FIG. 192. – View of station grounds and buildings, Fukuoka Experiment Station, Japan.

most strange was the fact that these stronger plants are never used in transplanting, as they do not thrive as well as those less vigorous.

Fig. 193. – Two Japanese ploughs.

We left the island of Kyushu in the evening of June 29th, crossing to the main island of Honshu, waiting in Shimonoseki for the morning train. The rice-planted valleys near Shimonoseki were relatively broad and the paddy fields had all been recently set in close rows about a foot apart and in hills in the rows. Mountain and hill lands were closely wooded, largely with coniferous trees about the base, but toward and at the summits they were green only with herbage cut for fertilizing and feeding stock. Many very small trees, often not more than 1 foot high, were growing on the recently cut-over areas; tall slender graceful bamboos clustered along the way and everywhere threw wonderful beauty into the landscape. Cartloads of their slender stems, 2 to 4 inches in diameter at the base and 20 or more feet long, were

FIG 194. – Plant breeding and variety test nursery rice plats, at Fukuoka Experiment Station.

moving along the generally excellent, narrow, seldom fenced roads. On the borders and pathways between rice fields many small stacks of straw were in waiting to be laid between the rows of transplanted rice, tramped beneath the water and overspread with mud to enrich the soil. The farmers here, as elsewhere, must contend against the scouring rush, varieties of grass and our common pigweeds, even in the rice fields. The large area of mountain and hill land and the relatively small area of cultivated land, not at this time under water and planted with rice, persisted throughout the journey.

If there could be any monotony for the traveller new to this land of beauty it must result from the quick shifting of scenes and in the way the landscapes are pieced together, outdoing the craziest patchwork woman ever attempted; the bits are almost never large; they are of every shape, even puckered and crumpled and tilted at all angles. Here is a bit of the journey: Beyond Habu the foothills are thickly wooded, largely with conifers. The valley is extremely narrow with only small areas for rice. Bamboo are growing in congenial places and we pass bundles of wood cut to stove length. Then we cross a long narrow valley practically all in rice, and then another not half a mile wide, just before reaching Asa. Beyond, the fields become limited in area with the bordering low hills recently cut over and a new growth springing up over them in the form of small shrubs among which are many pine. Now we are in a narrow valley between small rice fields or with none at all, but dash into one more nearly level, with wide areas in rice, just before reaching Onoda, at 10.30 a.m. and continuing three minutes' ride beyond, when we are again between hills without fields and where the trees are pine with clumps of bamboo. In four minutes more we are among small rice fields and at 10.35 have passed another gap and are crossing another valley checkered with rice fields and lotus ponds, but in one minute more the hills have closed in, leaving only room for the track. At 10.37 we are running along a narrow valley with its terraced rice fields where many of the hills show naked soil among the bamboo, scattering pine and other small trees; then we are out among garden patches thickly mulched with straw. At 10.38 we are between higher hills with but narrow areas for rice stretching close along the track, but in two minutes these are passed and we are among low hills with terraced dry fields. At 10.42 we are spinning along the level valley with its rice, but are quickly out again among hills with naked soil where erosion was marked. This is just before passing Funkai where we are following the course of a stream some 60 feet wide with but little cultivated land in small areas. At 10.47 we are again passing narrow rice fields near the track where the people are busy weeding with their hands, half knee-deep in water. At 10.53 we enter a broader valley stretching far to the south and seaward, but we had crossed it in one minute, shot through another gap, and at 10.55 are traversing a much broader valley largely given over to rice, but where some of the paddy fields are bearing matting rush set in rows and in hills after the manner of rice. It is here we pass Oyou and just beyond cross a stream confined between levees built some distance back from either bank. At 11.17 this plain is left and we enter a narrow valley without fields. Thus do most of the agricultural lands of Japan lie in the narrowest valleys, often steeply sloping, and into which jutting spurs create the greatest irregularity of boundary and slope.

The journey of this day covered 350 miles in fourteen hours, all of the way through a country of remarkable and peculiar beauty which can be duplicated nowhere outside the mountainous, rice-growing Orient and there only during fifteen days closing the transplanting season. There were neither high mountains nor broad valleys, no great rivers and but few lakes; neither rugged naked rocks, tall forest trees nor wide level fields reaching away to unbroken horizons. But the low, rounded, soil-mantled mountain tops clothed in herbaceous and young forest growth fell everywhere into lower hills and these into narrow steep valleys which dropped by a series of water-level benches, as seen in Fig. 195, to the main river courses. Each one of these millions of terraces, set about by its raised rim, was a silvery sheet of water dotted in the daintiest manner with bunches of rice just transplanted, not so close nor yet so high and over-spreading as to obscure the water, yet enough to impart to the surface a most delicate sheen of green. The grass-grown narrow rims retaining the water in the basins, cemented them into series of the most superb mosaics, shaped into the valley bottoms by artisan artists perhaps two thousand years before and maintained by their descendants through all the years since, that on them the rains and fertility from the mountains and the sunshine from heaven might be transformed by the rice plant into food for the families and support for the nation. Two weeks earlier the aspect of these landscapes was very different, and two weeks later the reflecting water would lie hidden beneath the growing and rapidly developing mantle of green, to go on changing until autumn, when all would be overspread with the ripened harvest of grain. What intensified the beauty of it all was the fact that only along the widest valley bottoms were the mosaics level, except the water surface of each individual unit, and these were always small. At one time we were riding along a descending series of steps and then along another rising through a winding valley to disappear around a projecting spur, and any-

FIG. 195. – Looking up a terraced valley between Hongo and Fukuyama, Japan.

where in the midst of it all might be standing Japanese cottages or villas with the water and the growing rice literally almost against the walls, as seen in Fig. 196, while a nearby high terrace might hold its water on a level with the roofs. Can one wonder that the Japanese love their country or that they are born and bred landscape artists.

Just before reaching Hongo there were considerable areas thrown into long narrow, much raised, east and west beds under

Fig. 196. – Group of houses standing in rice fields, on edge of terraces, surrounded by water.

covers of straw matting inclined at a slight angle toward the south, some 2 feet above the ground but open toward the north. What crop may have been grown here we did not learn, but the matting was apparently intended for shade, as it was hot midsummer weather, and we suspect it may have been ginseng. It was here, too, that we came into the region of the culture of matting rush, extensively grown in Hiroshima and Okayama prefectures, but less extensively all over the empire. As with rice, the rush is first grown in nursery beds from which it is transplanted to the paddies, one acre of nursery supplying sufficient stock for 10 acres of field. The plants are set twenty to thirty stalks in a hill in rows 7 inches apart with the hills 6 inches from centre to centre in the row. Very high fertilization is practised, costing from 120 to 240 yen per acre, or $60 to $120 annually, the fertilizer consisting of bean cake and plant ashes, or in recent years, sometimes of sulphate of ammonia for nitrogen, and superphosphate of lime. About 10 per cent of the amount of fertilizer required for the crop is applied at the time of fitting the ground, the balance being administered from time to time as the season advances. Two crops of the rush may be taken from the same

FIG. 197. – Fields of matting rush with recently transplanted rice, and Government salt fields in the background.

ground each year or it is grown in rotation with rice, but most extensively on the lands less readily drained and not so well suited for other crops. Fields of the rush, growing in alternation with rice, are seen in Fig. 38, and in Fig. 197, with the Government salt fields lying along the seashore beyond.

With the most vigorous growth the rush attain a height exceeding 3 feet and the market price varies materially with the length of the stems. Good yields, under the best culture, may be as high as 6·5 tons per acre of the dry stems but the average yield is less, that of 1905 being 8,531 pounds, for 9,655 acres. The value of the product ranges from $120 to $200 per acre.

It is from this material that mats are woven in standard sizes, to be laid over padding, upholstering the floors which are the seats of all classes in Japan, used in the manner seen in Fig. 198.

FIG. 198. – Group of Japanese girls playing the game of flower cards, in the usual attitude of sitting on the matting-covered floor.

'Two little maids I've heard of, each with a pretty taste,
Who had two little rooms to fix and not an hour to waste.
Eight thousand miles apart they lived, yet on the selfsame day
The one in Nikko's narrow streets, the other on Broadway,
They started out, each happy maid her heart's desire to find,
And her own dear room to furnish just according to her mind.

When Alice went a-shopping, she bought a bed of brass,
A bureau and some chairs and things and such a lovely glass
To reflect her little figure – with two candle brackets near –
And a little dressing-table that she said was simply dear!
A bookshelf low to hold her books, a little china rack,
And then, of course, a bureau set and lots of bric-à-brac;

A dainty little escritoire, with fixings all her own;
And just for her convenience, too, a little telephone.
Some oriental rugs she got, and curtains of madras,
With 'cunning' ones of lace inside, to go against the glass;
And then a couch, a lovely one, with cushions soft to crush,
And forty pillows, more or less, of linen, silk and plush;
Of all the ornaments besides I couldn't tell the half,
But wherever there was nothing else, she stuck a photograph.
And then, when all was finished, she sighed a little sigh,
And looked about with just a shade of sadness in her eye:
"For it needs a statuette or so – a fern – a silver stork –
Oh, something, just to fill it up!" said Alice of New York.

When little Oumi of Japan went shopping, pit-a-pat,
She bought a fan of paper and a little sleeping mat;
She set beside the window a lily in a vase,
And looked about with more than doubt upon her pretty face:
"For, really – don't you think so? – with the lily and the fan,
It's a little overcrowded!" said Oumi of Japan.'

(*Margaret Johnson in St. Nicholas Magazine.*)

In the rural homes of Japan during 1906 there were woven 14,497,058 sheets of these floor mats and 6,628,772 sheets of other matting, having a combined value of $2,815,040, and in addition, from the best quality of rush grown upon the same ground, aggregating 7,657 acres that year, there were manufactured for the export trade, fancy mattings having the value of $2,274,131. Here is a total value, for the product of the soil and for the labour put into the manufacture, amounting to $664 per acre for the area named.

At the Akashi agricultural experiment station, under the Directorship of Professor Ono, we saw some of the methods of fruit culture as practised in Japan. He was conducting experiments with the object of improving methods of heading and training pear trees, to which reference was made on page 32. A study was also being made of the advantages and disadvantages associated with covering the fruit with paper bags, examples of which are seen in Figs. 5 and 6. The bags were being made at the time of our visit, from old newspapers cut, folded and pasted by women. Naked cultivation was practised in the orchard, and fertilizers consisting of fish guano and superphosphate of lime were being applied twice each year in amounts aggregating a cost of $24 per acre.

Pear orchards of native varieties, in good bearing, yield returns of 150 yen per tan, and those of European varieties, 200 yen per tan, which is at the rate of $300 and $400 per acre. The bibo so extensively grown in China was being cultivated here also and was yielding about $320 per acre.

It was here that we first met the cultivation of a variety of burdock grown from the seed, three crops being taken each season where the climate is favourable, or as one of three in the multiple crop system. It is grown for the root, yielding a crop valued at $40 to $50 per acre. One crop, planted in March, was being harvested July 1st.

During our ride to Akashi on the early morning train we passed long processions of carts drawn by cattle, horses or by men, moving along the country road which paralleled the railway, all loaded with the waste of the city of Kobe, going to its destination in the fields, some of it a distance of 12 miles, where it was sold at from 54 cents to $1.63 per ton.

At several places along our route from Shimonoseki to Osaka we had observed the application of slacked lime to the water of the rice fields, but in this prefecture, Hyogo, where the station is located, its use was prohibited in 1901, except under the direction of the station authorities, where the soil was acid or where it was needed on account of insect troubles. Up to this time it had been the custom of farmers to apply slacked lime at the rate of 3 to 5 tons per acre, paying for it $4.84 per ton. The first restrictive legislation permitted the use of 82 pounds of lime with each 827 pounds of organic manure, but as the farmers persisted in using much larger quantities, complete prohibition was resorted to.

Reference has been made to subsidies encouraging the use of composts, and in this prefecture prizes are awarded for the best compost heaps in each county, examinations being made by a committee. The composts receiving the four highest awards in each county are allowed to compete with those in other counties for a prefectural prize awarded by another committee.

The 'pink clover' grown in Hyogo after rice, as a green manure crop, yields under favourable conditions 20 tons of the green product per acre, and is usually applied to about three times the area upon which it grew, at the rate of 6·6 tons per acre, the stubble and roots serving for the ground upon which the crop grew.

On July 3rd we left Osaka, going south through Sakai to Wakayama, thence east and north to the Nara Experiment Station. After passing the first two stations the route lay through a very flat, highly cultivated garden section with cucumbers trained on trellises, many squash in full bloom, with fields of taro, ginger and many other vegetables. Beyond Hamadera considerable areas of flat sandy land had been set close with pine, but with intervening areas in rice, where the growers were using the revolving weeder seen in Fig. 12. At Otsu broad areas are in rice but

FIG. 199. – Distribution of old stubble and the working of it beneath the water and mud to serve as fertilizer.

here worked with the short-handled claw weeders, and stubble from a former crop had been drawn together into small piles, seen in Fig. 199, which later would be carefully distributed and worked beneath the mud.

Much of the mountain lands in this region, growing pine, is owned by private parties and the growth is cut at intervals of ten, twenty or twenty-five years, being sold on the ground to

Fig. 200. – Irrigating with the Japanese circumferential foot-power water wheel, near Hashimoto.

those who will come and cut it at a price of 40 sen for a one-horse load, as already described (page 142).

The course from here was up the rather rapidly rising Kiigawa valley where much water was being applied to the rice fields by various methods of pumping, among them numerous current wheels; an occasional power-pump driven by cattle; and very commonly the foot-power wheel where the man walks on the circumference, steadying himself with a long pole, as seen in the field, Fig. 200. It was here that a considerable section of the hill slope had been very recently cut over, the area showing light in the engraving. It was in the vicinity of Hashimoto on this route, too, that the two beautiful views reproduced in Figs. 134 and 135 were taken.

At the experiment station it was learned that within the prefecture of Nara, having a population of 558,314, two-thirds of the 107,574 acres of cultivated land was in rice. Within the province there are also about one thousand irrigation reservoirs with an average depth of 8 feet. The rice fields receive 16·32 inches of irrigation water in addition to the rain.

Of the uncultivated hill lands, some 2,500 acres contribute green manure for fertilization of fields. Reference has been made to the production of compost for fertilizers on page 186. The amount recommended in this prefecture as a yearly application for two crops grown is:

Organic matter	3,711 to 4,640 lbs. per acre
Nitrogen	105 to 131 lbs. per acre
Phosphorus	35 to 44 lbs. per acre
Potassium	56 to 70 lbs. per acre

These amounts, on the basis of the table, page 189, are nearly sufficient for a crop of 30 bushels of wheat, followed by one of 30 bushels of rice, the phosphorus being in excess and the potassium not quite enough, supposing none to be derived from other sources.

At the Nara hotel, one of the beautiful Japanese inns where we stopped, our room opened upon a second story veranda from which one looked down upon a beautiful, tiny lakelet, some 20 by 80 feet, within a diminutive park scarcely more than 100 by 200 feet. The lakelet had its grassy rocky banks overhung with trees and shrubs planted in all the wild disorder and beauty of nature; bamboo, willow, fir, pine, cedar, red-leaved maple, catalpa, with other kinds, and through these, along the shore, wound a woodsy, well trodden, narrow footpath leading from the inn to a half-hidden cottage, apparently quarters for the maids, as they were frequently passing to and fro. A suggestion of how such wild beauty is brought right to the very doors in Japan may be gained from Fig. 201, which is an instance of parking effect on a still smaller scale than that described.

On the morning of July 6th, with two men for each of our rickshas, we left the Yaami hotel for the Kyoto Experiment

Station, some 2 miles to the south-west of the city limits. As soon as we had entered upon the country road we found ourselves in a procession of cart men each drawing a load of six large covered receptacles of about 10 gallons capacity, and filled with the city's waste. Before reaching the station we had passed fifty-two of these loads. On our return the procession was still moving in the same direction and we passed sixty-one others, so that during at least five hours there had moved over this section of road leading

Fig. 201. – Beauty at home in Japan.

into the country, away from the city, not less than 90 tons of waste. Along other roadways similar loads were moving. These freight carts and those drawn by horses and bullocks were all provided with long racks, and when the load is not sufficient to cover the full length it is always divided equally and placed near each end, thus taking advantage of the elasticity of the body to give the effect of springs, lessening the draught and the wear and tear.

One of the most common commodities coming into the city

along the country roads was fuel from the hill lands, in split sticks tied in bundles; as bundles of limbs 24 to 30 inches, and sometimes 4 to 6 feet, long; and in the form of charcoal made from trunks and stems $1\frac{1}{2}$ to 6 inches long, and baled in straw matting. Most of the draught animals used in Japan are either cows, bulls or stallions; at least we saw very few oxen and few geldings.

As early as 1895 the Government began definite steps looking to the improvement of horse breeding, appointing at that time a commission to devise comprehensive plans. This led to progressive steps finally culminating in 1906 in the Horse Administration Bureau, whose duties were to extend over a period of thirty years, divided into two intervals, the first eighteen and the second twelve years. During the first interval it is contemplated that the Government will acquire 1,500 stallions to be distributed throughout the country for the use of private individuals, and during the second period it is the expectation that the system will have completely renovated the stock and familiarized the people with proper methods of management so that matters may be left in their hands.

As our main purpose and limited time required undivided attention to agricultural matters, and of these to the long established practices of the people, we could give but little time to sight-seeing or even to a study of the efforts being made for the introduction of improved agricultural methods and practices. But in the very old city of Kyoto, which was the seat of the Mikado's court from before A.D. 800 until 1868, we did pay a short visit to the Kiyomizu temple, situated some 300 yards south from the Yaami hotel. It faces the Maruyaami park with its centuries-old giant cherry tree, having a trunk of more than 4 feet through and wide-spreading branches, now much propped up to guard against accident. These cherry trees are very extensively used for ornamental purposes in Japan with striking effect. The tree does not produce an edible fruit, but is very beautiful when in full bloom, as may be seen from Fig. 202. It was these trees that were sent by the Japanese government to the United States for use at Washington, but the first lot were destroyed because they were found to be infested and threatened danger to native trees.

Kyoto stands amid surroundings of wonderful beauty, the site apparently having been selected with rare acumen for its possibilities in large landscape effects, and these have been developed with that fullness and richness which the greatest artists might be content to approach. We are thinking particularly of the Kiyomizu-dera, or rather of the marvellous beauty of tree and foliage which has overgrown it and swept far up and over the mountain summit, leaving the temple half hidden at the base. No words, no brush, no photographic art can transfer the effect. One must see to feel the influence for which it was created, and

FIG. 202. – Admiring cherry blossoms.

scores of people, very old and very young, nearly all Japanese, and more of them on that day from the poorer rather than from the well-to-do class, were there, all withdrawing reluctantly, like ourselves, looking backward, under the spell. So potent and impressive was that something from the great overshadowing beauty of the mountain, that all along up the narrow, shop-lined street leading to the gateway of the temple, the tiniest bits of park effect were flourishing in the most impossible situations; and as Professor Tokito and myself were coming away we chanced upon six little roughly dressed lads laying out in the sand an elaborate little park, quite 9 by 12 feet. They must have been at it hours, for there were ponds, bridges, tiny hills and ravines and much planting in moss and other little greens. So intent on their task were they that we stood watching full two minutes before our presence attracted their attention, and yet the oldest of the group must have been under ten years of age.

Within the temple, as the peasant men and women came before the shrine and grasped the long depending rope knocker, with the heavy knot in front of the great gong, swinging it to strike three rings, announcing their presence before their God, then kneeling to offer prayers, one could not fail to realize the deep sincerity and faith expressed in face and manner, while they were oblivious to all else. No Christian was ever more devout and one may well doubt if any ever arose from prayer more uplifted than these. Who need believe they did not look beyond the imagery and commune with the Eternal Spirit?

When returning to the city from the Kyoto Experiment Station several fields of Japanese indigo were passed, growing in water under the conditions of ordinary rice culture, Fig. 203 being a view of one of these. The plant is *Poligonum tinctoria*, a close relative of the smartweed. Before the importation of aniline and alizarin dyes, which amounted in 1907 to 160,558 pounds and 7,170,320 pounds respectively, the cultivation of indigo was much more extensive than at present, amounting in 1897 to 160,460,000 pounds of the dried leaves; but in 1906 the production had fallen to 58,696,000 pounds, 45 per cent of which was grown in the prefecture of Tokushima in the eastern part of the island of Shikoku. The population of this prefecture is 707,565, or 4·4 people to each of the 159,450 acres of cultivated field, and yet 19,969 of these acres bore the indigo crop, leaving more than five people to each food-producing acre.

The plants for this crop are started in nursery beds in February and transplanted in May, the first crop being cut the last of June or first of July, when the fields are again fertilized, the stubble throwing out new shoots and yielding a second cutting the last of August or early September. A crop of barley may have preceded one of indigo, or the indigo may be set following a crop of rice. Such practice, with the high fertilization for every crop, goes a long way toward supplying the necessary food. The dense population, too, has permitted the manufacture of the indigo as a home industry among the farmers, enabling them to exchange the spare labour of the family for cash. The manufactured product from the reduced planting in 1907 was worth $1,304,610, 45 per cent of which was the output of the rural population of the prefecture of Tokushima, which they could exchange for rice and other necessaries. The land in rice in this prefecture in 1907 was 73,816 acres, yielding 114,380,000 pounds, or more than 161 pounds

Fig. 203. – Field of Japanese indigo, just outside the city of Kyoto.

to each man, woman and child, and there were 65,665 acres bearing other crops. Besides this there are 874,208 acres of mountain and hill land in the prefecture which supply fuel, fuel ashes and green manure for fertilizer; run-off water for irrigation; lumber and remunerative employment for service not needed in the fields.

The journey was continued from Kyoto July 7th, by the route leading north-eastward, skirting Lake Biwa which we came upon suddenly on emerging from a tunnel as the train left Otani. At many places we passed water-wheels, busily turning, and usually 12 to 16 feet in diameter but oftenest only as many inches thick. Until we had reached Lake Biwa the valleys were narrow with only small areas in rice. Tea plantations were common on the higher cultivated slopes, as well as gardens on the terraced hill-sides growing vegetables of many kinds. Often the ground was heavily mulched with straw, while the wooded or grass-covered slopes still further up showed the usual systematic periodic cutting. After passing the west end of the lake, rice fields were nearly continuous and extensive. Before reaching Hachiman we crossed a stream leading into the lake but confined between levees more than 12 feet high, and we had already passed beneath two raised viaducts after leaving Kusatsu. Other crops were being grown side by side with the rice on similar lands and apparently in rotation with it, but on sharp, narrow, close ridges 12 to 14 inches high. As we passed eastward we entered one of the important mulberry districts where the fields are graded to two levels, the higher occupied with mulberry or other crops not requiring irrigation, while the lower was devoted to rice or crops grown in rotation with it.

On the Kisogawa, at the station of the same name, there were four anchored floating water-power mills propelled by two pair of large current wheels stationed fore and aft, each pair working on a common axle from opposite sides of the mill, driven by the force of the current.

At Kisogawa we had entered the northern end of one of the largest plains of Japan, some 30 miles wide and extending 40 miles southward to Owari bay. The plain has been extensively graded to two levels, the benches being usually not more than 2 feet above the paddy fields, and devoted to various dry land crops, including the mulberry. The soil is decidedly sandy in character, but the mean yield of rice for the prefecture is 37 bushels per acre and above the average for the country at large. An analysis of the soils at the sub-experiment station north of Nagoya shows the following content of the three main plant food elements:

	Nitrogen	Phosphorus	Potassium
		Pounds per million	
		In paddy field	
Soil	1,520	769	805
Subsoil	810	756	888
		In upland field	
Soil	1,060	686	1,162
Subsoil	510	673	1,204

The green manure crops on this plain are chiefly two varieties of the 'pink clover,' one sown in the fall and one about May 15th, the first yielding as high as 16 tons green weight per acre and the other from 5 to 8 tons.

On the plain distant from the mountain and hill land the stems of agricultural crops are largely used as fuel and the fuel ashes are applied to the fields at the rate of 10 kan per tan, or 330 pounds per acre, worth $1.20, little lime, as such, being used.

In the prefecture of Aichi, largely in this plain, with an area of cultivated land equal to about 16 of our government townships, there is a population of 1,752,042, or a density of 4·7 per acre, and the number of households of farmers was placed at 211,033, thus giving to each farmer's family an average of 1·75 acres, their chief industries being rice and silk culture.

Soon after leaving the Agricultural Experiment Station of Aichi prefecture at An Jo we crossed the large Yahagigawa, flowing between strong levees above the level of the rice fields. Mulberries, with burdock and other vegetables, were growing upon all the tables raised one to two feet above the rice fields, and these features continued past Okasaki, Koda and Kamagori, where the hills in many places had been recently cut clean of the low forest growth and where we passed many large stacks of pine boughs tied in bundles for fuel. After passing Goyu 65 miles east from Nagoya, mulberry was the chief crop. Then came a plain country which had been graded and levelled at great cost of labour, the benches with their square shoulders standing 3 to 4 feet above the paddy fields; and after passing Toyohashi some distance we were surprised to cross a rather wide section of comparatively level land overgrown with pine and herbaceous plants which had evidently been cut and recut many times. Beyond Futagawa rice fields were laid out on what appeared to be similar land but with soil a little finer in texture, and still further along were other flat areas not cultivated.

At Maisaka quite half the cultivated fields appear to be in mulberry with ponds of lotus plants in low places, while at Hamamatsu the rice fields are interspersed with many square-shouldered tables raised 3 to 4 feet and occupied with mulberry or vegetables. As we passed upon the flood plain of the Tenryugawa, with its nearly dry bed of coarse gravel half a mile wide, the dwellings of farm villages were many of them surrounded with nearly solid, flat-topped, trimmed evergreen hedges 9 to 12 feet high, of the umbrella pine, forming beautiful and effective screens.

At Nakaidzumi we had left the mulberry orchards for those of tea, rice still holding wherever paddy fields could be formed. Here, too, we met the first fields of tobacco, and at Fukuroi and Homouchi large quantities of imported Manchurian bean cake were stacked about the station, having evidently been brought by rail. At Kanaya we passed through a long tunnel and were in the valley of the Oigawa, crossing the broad, nearly dry stream over a bridge of nineteen long spans and were then in the prefecture of Shizuoka where large fields of tea spread far up the hillsides, covering extensive areas, but after passing the next station, and for 17 miles before reaching Shizuoka, we traversed a level stretch of nearly continuous rice fields.

The Shizuoka Experiment Station is devoting special attention to the interests of horticulture, and progress has already been made in introducing new fruits of better quality and in improving the native varieties. The native pears and peaches, as we found them served on the hotel tables in either China or Japan, were not particularly attractive in either texture or flavour, but we were here permitted to test samples of three varieties of ripe figs of fine flavour and texture, one of them as large as a good-sized pear. Three varieties of fine peaches were also shown, one unusually large and with delicate deep rose tint, including the flesh. If such peaches could be canned so as to retain their delicate colour they would prove very attractive for the table. The flavour and texture of this peach were also excellent, as was the case with two varieties of pears.

The station was also experimenting with the production of marmalades and we tasted three very excellent brands, two of them lacking the bitter flavour. It would appear that in Japan, Korea and China there should be a very bright future along the lines of horticultural development, leading to the utilization of the extensive hill lands of these countries and the development of a very extensive export trade, both in fresh fruits and marmalades, preserves and the canned forms. They have favourable climatic and soil conditions and great numbers of people with temperament and habits well suited to the industries, as well as an enormous home need which should be met, in addition to the large possibilities in the direction of a most profitable export trade which would increase opportunities for labour and bring needed revenue to the people. In Fig. 204 are three views at this station, the lower showing a steep terraced hillside set with oranges and other fruits, holding out a bright promise for the future.

Peach orchards were here set on the hill lands, the trees 6 feet apart each way. They come into bearing in three years, remain productive ten to fifteen years, and the returns are 50 to 60 yen per tan, or at the rate of $100 to $120 per acre. The usual fertilizers for a peach orchard are the manure-earth-compost, applied at the rate of 3,300 pounds per acre, and fish guano applied in rotation and at the same rate.

Shizuoka is one of the large prefectures, having a total area of 3,029 square miles; 2,090 of which are in forest, 438 in pasture and *genya* land, and 501 square miles cultivated, not quite one-half of which is in paddy fields. The mean yield of rice is nearly 33 bushels per acre. The prefecture has a population of 1,293,470, or about four to the acre of cultivated field, and the total crop of rice is such as to provide 236 pounds to each person.

At many places along the way as we left Shizuoka July 10th for Tokyo, farmers were sowing broadcast, on the water, over their rice fields, some pulverized fertilizer, possibly bean cake. Near the railway station of Fuji, and after crossing the boulder gravel bed of the Fujikawa which was a full quarter of a mile wide, we were traversing a broad plain of paddy fields with their raised tables, but on them pear orchards were growing, trained to their overhead trellises. About Suduzuka grass was being cut with sickles along the canal dikes for use as green manure in the rice fields, which stretched eastward more than 6 miles to beyond Hara. Here we passed into a tract of dry land crops consisting of mulberry, tea and various vegetables, with more or less of dry land rice, but we returned to the paddy land again at Numazu, in another 4 miles.

It was at this station that the railway turns northward to skirt the eastern flank of the beautiful Fuji-yama, rising to higher lands of a brown loamy character, showing many large boulders 2 feet in diameter. Horses were here moving along the roadways under large saddle loads of green grass, going to the paddy fields

FIG. 204. – Views of buildings and grounds at the Shizuoka Experiment Station.

from the hills, which in this section are quite free from all but herbaceous growth, well covered and green. Considerable areas were growing maize and buckwheat, the latter being ground into flour and made into macaroni which is eaten with chopsticks, and used to give variety to the diet of rice and naked barley. At Gotenba, where tourists leave the train to ascend Fuji-yama, the road turns eastward again and descends rapidly through many tunnels, crossing the wide gravelly channel of the Sakawagawa. The river was then carrying but little water, like all the other main streams we had crossed, although we were in the rainy season. This was partly because the season was yet not far advanced; partly because so much water was being taken upon the rice fields, and again because the drainage is so rapid down the steep slopes and comparatively short water courses. Beyond Yamakita the railway again led along a broad plain set in rice and the hill slopes were terraced and cultivated nearly to their summits.

Swinging strongly south-eastward, the coast was reached at Noduz in a hilly country producing chiefly vegetables, mulberry and tobacco, the latter crop being extensively grown eastward nearly to Oiso, beyond which, after a mile of sweet potatoes, squash and cucumbers, there were paddy fields of rice in a flat plain. Before Hiratsuka was reached the paddy fields were left and the train was crossing a comparatively flat country with a sandy, sometimes gravelly, soil where mulberries, peaches, egg-plants, sweet potatoes and dry land rice were interspersed with areas still occupied with small pine and herbaceous growth, or where small pine had been recently set. Similar conditions prevailed after we had crossed the broad channel of the Banyugawa and well toward and beyond Fujishiwa where a levelled plain has its tables scattered among the fields of paddy rice. This is the south-west margin of the Tokyo plain, the largest in Japan, lying in five prefectures, whose aggregate area of 1,739,200 acres of arable lands was worked by 657,235 families of farmers; 661,613 acres of which was in paddy rice, producing annually some 19,198,000 bushels, or 161 pounds for each of the 7,194,045 men, women and children in the five prefectures, 1,818,655 of whom were in the capital city, Tokyo.

Three views taken in the eastern portion of this plain in the prefecture of Chiba, July 17th, are seen in Fig. 205, in two of which

Fig. 205. – Three landscapes in the Tokyo plain, the upper two largely in sweet potatoes, following wheat, the lower in peanuts.

shocks of wheat were still standing among the growing crops, badly weathered and the grain sprouting as the result of the rainy season. Peanuts, sweet potatoes and millet were the main dry land crops then on the ground, with paddy rice in the flooded basins. Windsor beans, rape, wheat and barley had been harvested. One family with whom we talked were threshing their wheat. The crop had been a good one and was yielding between 38·5 and 41·3 bushels per acre, worth at the time $35 to $40. On the same land this farmer secures a yield of 352 to 361 bushels of potatoes, which at the market price at that time would give a gross earning of $64 to $66 per acre.

Reference has been made to the extensive use of straw in the cultural methods of the Japanese. This is notably the case in their truck garden work, and two phases of this are shown in Fig. 206. In the lower section of the illustration the garden has been ridged and furrowed for transplanting, the sets have been laid and the roots covered with a little soil; then, in the middle section, showing the next step in the method, a layer of straw has been pressed firmly above the roots, and in the final step this would be covered with earth. Adopting this method the straw is so placed that (1) it acts as an effective mulch without in any way interfering with the capillary rise of water to the roots of the sets; (2) it gives deep, thorough aeration of the soil, at the same time allowing rains to penetrate quickly, drawing the air after it; (3) the ash ingredients carried in the straw are leached directly to the roots where they are needed; (4) and finally the straw and soil constitute a compost where the rapid decay liberates plant food gradually and in the place where it will be most readily available. The upper section of the illustration shows rows of egg-plants very heavily mulched with coarse straw, the quantity being sufficient to act as a most effective mulch, to largely prevent the development of weeds and to serve during the rainy season as a very material fertilizer.

In growing such dry land crops as barley, beans, buckwheat or dry land rice the soil of the field is at first fitted by ploughing or spading, then furrowed deeply where the rows are to be planted. Into these furrows fertilizer is placed and covered with a layer of earth upon which the seed is planted. When the crop is up, if a second fertilization is desired, a furrow may be made alongside

FIG. 206. – Two methods of utilizing coarse straw and litter for mulching and fertilizing at the same time.

each row, into which the fertilizer is sowed and then covered. When the crop is so far matured that a second may be planted, a new furrow is made, either midway between two others or adjacent to one of them, fertilizer applied and covered with a layer of soil and the seed planted. In this way the least time possible is lost during the growing season, all the soil of the field doing duty in crop production.

It was our privilege to visit the Imperial Agricultural Experiment Station at Nishigahara, near Tokyo, which is charged with the leadership of the general and technical agricultural research work for the Empire. The work is divided into the sections of agriculture, agricultural chemistry, entomology, vegetable pathology, tobacco, horticulture, stock breeding, soils, and tea manufacture, each with their laboratory equipment and research staff, while the forty-one prefectural stations and fourteen sub-stations are charged with the duty of handling all specific local, practical problems and with testing out and applying conclusions and methods suggested by the results obtained at the central station, together with the local dissemination of knowledge among the farmers of the respective prefectures.

A comprehensive soil survey of the arable lands of the Empire has been in progress since before 1893, excellent maps being issued on a scale of 1 to 100,000, or about 1·57 in. to the mile, showing the geological formations in eight colours with subdivisions indicated by letters. Some eleven soil types are recognized, based on physical composition, and the areas occupied by these are shown by means of lines and dots in black printed over the colours. Typical profiles of the soil to depths of three metres are printed as insets on each sheet and localities where these apply are indicated by corresponding numbers in red on the map.

Elaborate chemical and physical studies are also being made in the laboratories of samples of both soil and subsoil. The Imperial Agricultural Experiment Station is well equipped for investigation work along many lines and that for soils is notably strong. In Fig. 207 may be seen a portion of the large immersed cylinders which are filled with typical soils from different parts of the Empire, and Fig. 208 shows a portion of another part of their elaborate outfit for soil studies which are in progress.

It is found that nearly all cultivated soils of Japan are acid to litmus, and this they are inclined to attribute to the presence of acid hydro-aluminum silicates.

The Island Empire of Japan stretches along the Asiatic coast through more than 29° of latitude from the southern extremity of Formosa northward to the middle of Saghalin, some 2,300 statute miles; or from the latitude of middle Cuba to that of north Newfoundland and Winnipeg; but the total land area is only 175,428 square miles, and less than that of the three states

FIG. 207. – Section of soil study field, Imperial Agricultural Experiment Station, Tokyo.

of Wisconsin, Iowa and Minnesota. Of this total land area only 23,698 square miles are at present cultivated; 7,151 square miles in the three main islands are weed and pasture land. Less than 14 per cent of the entire land area is at present under cultivation.

If all lands having a slope of less than 15° may be tilled, there yet remain in the four main islands, 15,400 square miles to bring under cultivation, which is an addition of 65·4 per cent to the land already cultivated.

In 1907 there were in the Empire some 5,814,362 households

Fig. 208. – Part of equipment for chemical soil studies, Imperial Agricultural Experiment Station, Tokyo.

of farmers tilling 15,201,969 acres and feeding 3,522,877 additional households, or 51,742,398 people. This is an average of 3·4 people to the acre of cultivated land, each farmer's household tilling an average of 2·6 acres.

The lands yet to be reclaimed are being put under cultivation rapidly, the amount improved in 1907 being 64,448 acres. If the new lands to be reclaimed can be made as productive as those now in use there should be opportunity for an increase in population to the extent of about 35,000,000 without changing the present ratio of 3·4 people to the acre of cultivated land.

While the remaining lands to be reclaimed are not as inherently productive as those now in use, improvements in management will more than compensate for this, and the Empire is certain to double its present maintenance capacity and provide for at least 100 million people with many more comforts of home and more satisfaction for the common people than they now enjoy.

Since 1872 there has been an increase in the population of Japan amounting to an annual average of about 1·1 per cent, and if this rate is maintained the 100 million mark would be passed in less than sixty years. It appears probable, however, that the increased acreage put under cultivation and pasturage combined, will more than keep pace with the population up to this limit, while the improvement in methods and crops will readily permit a second like increment to her population, bringing that for the present Empire up to 150 millions. Against this view, perhaps, is the fact that the rice crop of the twenty years ending in 1906 is only 33 per cent greater than the crop of 1838.

In Japan, as in the United States, there has been a strong movement from the country to the city as a natural result of the large increase in manufactures and commerce, and the small amount of land per each farmer's household. In 1903 only ·23 per cent of the population of Japan were living in villages of less than 500, while 79·06 per cent were in towns and villages of less than 10,000 people, 20·7 per cent living in those larger. But in 1894 84·36 per cent of the population were living in towns and villages of less than 10,000, and only 15·64 per cent were in cities, towns and villages of over 10,000 people; and while during these ten years the rural population had increased at the rate of 640 per 10,000, in cities the increase had been 6,174 per 10,000.

Japan has been and still is essentially an agricultural nation and in 1906 there were 3,872,105 farmers' households, whose chief work was farming, and 1,581,204 others whose subsidiary work was farming, or 60·2 per cent of the entire number of households. A like ratio holds in Formosa. Wealthy landowners who do not till their own fields are not included.

Of the farmers in Japan some 3·334 per cent own and work their land. Those having smaller holdings, who rent additional land, make up 46·03 per cent of the total farmers; while 20·63 per cent are tenants who work 44·1 per cent of the land. In 1892 only 1 per cent of the land holders owned more than 25 acres each; those holding between 25 acres and 5 acres made up 11·7 per cent; while 87·3 per cent held less than 5 acres each. A man owning 75 acres of land in Japan is counted among the 'great land-holders.' It is never true, however, except in the Hokkaido, which is a new country agriculturally, that such holdings lie in one body.

Statistics published in *Agriculture in Japan*, by the Agricultural Bureau, Department of Agriculture and Commerce, permit the following statements of rent, crop returns, taxes and expenses to be made. The wealthy landowners who rent their lands receive returns like these:

	For paddy field, per acre.	For upland field, per acre.
Rent	$27.98	$13.53
Taxes	7.34	1.98
Expenses	1.72	2.48
Total expenses	$9.06	$4.46
Net profit	18.92	9.07

It is stated, in connection with these statistics, that the rate of profit for land capital is 5·6 per cent for the paddy field, and 5·7 per cent for the upland field. This makes the valuation of the land about $338 and $159 per acre, respectively. A land holder who owns and rents 10 acres of paddy field and 10 acres of upland field would, at these rates, realize a net annual income of $279.90.

Peasant farmers who own and work their lands receive per acre an income as follows:

	For paddy field, per acre.	For upland field, per acre.
Crop returns	$55.00	$30.72
Taxes	7.34	1.98
Labour and expenses	36.20	24.00
Total expenses	$43.54	$25.98
Net profit	11.46	4.74

The peasant farmer who owns and works 5 acres, 2·5 of paddy and 2·5 of upland field, would realize a total net income of $40.50. This is after deducting the price of his labour. With that included, his income would be something like $91.

Tenant farmers who work some 41 per cent of the farm lands of Japan would have accounts something as follows:

	For paddy field, 1 crop. per acre.	For paddy field, 2 crops. per acre.	For upland field, per acre.
Crop returns	$49.03	$78.62	$41.36
Tenant fee	23.89	31.58	13.52
Labour	15.78	25.79	14.69
Fertilization	7.82	17.30	10.22
Seed	.82	1.40	1.57
Other expenses	1.69	2.82	1.66
Total expenses	$50.00	$78.89	$41.66
Net profit	—.97	—.27	—.30

This statement indicates that tenant farmers do not realize enough from the crops to quite cover expenses and the price named for their labour. If the tenant were renting 5 acres, equally divided between paddy and upland field, the earning would be $7.300 or $99.73 according as one or two crops are taken from the paddy field. This represents what he realizes on his labour, his other expenses absorbing the balance of the crop value.

But the average area tilled by each Japanese farmer's household is only 2.6 acres, hence the average earning of the tenant household would be $37.95 or $51.86. A clearer view of the difference in the present condition of farmers in Japan and of those in the

United States may be gained by making the Japanese statement on the basis of our 160-acre farm, as expressed in the table below:

	For paddy field. For 80 acres.	For upland field. For 80 acres.	Total 160 acres.
Crop returns	$1,400.00	$2,457.60	$6,857.60
Taxes	$587.20	$158.40	$745.60
Expenses	1,633.60	744 80	2,378.40
Labour	1,262.40	1,175.20	2,437.60
Total cost	$3,488.20	$2,078.40	$5,561.60
Net return	916.80	379 20	1,296.00
Return including labour	2,179.20	1,551.40	3,733.60

In the United States the 160-acre farm is managed by and supports a single family, but in Japan, as the average household works but 2.6 acres, the earnings of the 160 acres are distributed among some 61 households, making the net return to each but $21.25, instead of $1,296, and including the labour as earning, the income would be $39.96 more, or $60.67 per household instead of $3,733.60, the total for a 160-acre farm worked under Japanese conditions.

These figures reveal something of the tense strain and of the terrible burden which is being carried by these people, over and above that required for the maintenance of the household. The tenant who raises one crop of rice pays a rental of $23.89 per acre. If he raises two crops he pays $31.58; if it is upland field, he pays $13.52. To these amounts he adds $10.33, $21.52 or $13.45 respectively for fertilizer, seed and other expenses, making a total investment of $34.22, $53.10 or $26 97 per acre, which would require as many bushels of wheat sold at a dollar a bushel to cover this cost. In addition to this he assumes all the risks of loss from weather, from insects and from blight, in the hope that he may recoup his expenses and in addition have for his services $14.81, $25 52 or $14.39 for the season's work.

The burdens of society, which have been and still are so largely burdens of war and of government, with all nations, are reflected with almost blinding effect in the land taxes of Japan, which range from $1.98 on the upland, to $7.34 per acre on the paddy fields, making a quarter section, without buildings, carry a burden of $300 to $1,100 annually. Japan's budget in 1907 was $134,941,113, which is at the rate of $2.60 for each man, woman and child; $8.90 for each acre of cultivated land, and $23 for each household in the Empire. When such is the case it is not strange that scenes like Fig. 209 are common in Japan to-day where, after seventy years, toil may not cease.

FIG. 209. – After seventy years, toil may not cease.

There is a bright, as well as a pathetic side to scenes like this. The two have shared for fifty years, but if the days have been full of toil, with them have come strength of body, of mind and sterling character. If the burdens have been heavy, each has made the other's lighter, the satisfaction fuller, the joys keener, the sorrows less difficult to bear; and the children who came into the home and have gone from it to perpetuate new ones, could not well be other than such as to contribute to the foundations of nations of great strength and long endurance.

Reference has been made to the large amount of work carried on in the farmers' households by the women and children, and by the men when they are not otherwise employed. The earnings of this subsidiary work have materially helped to piece out the meagre income and to meet the relatively high taxes and rent.

INDEX

D

E

F

G

K

L

M

T